Sustainable Building and Built Environments to Mitigate Climate Change in the Tropics

Tri Harso Karyono • Robert Vale • Brenda Vale
Editors

Sustainable Building and Built Environments to Mitigate Climate Change in the Tropics

Conceptual and Practical Approaches

Springer

Editors
Tri Harso Karyono
School of Architecture
Tanri Abeng University
Jakarta, Indonesia

Robert Vale
School of Architecture and Design
Victoria University of Wellington
Wellington, New Zealand

Brenda Vale
School of Architecture and Design
Victoria University of Wellington
Wellington, New Zealand

ISBN 978-3-319-49600-9 ISBN 978-3-319-49601-6 (eBook)
DOI 10.1007/978-3-319-49601-6

Library of Congress Control Number: 2017932529

Printed on acid-free paper

This Springer imprint is published by Springer Nature
The registered company is Springer International Publishing AG
The registered company address is: Gewerbestrasse 11, 6330 Cham, Switzerland

Preface

This book is about presenting solutions to problems of the built environment due to the current changes of the world climate. A number of efforts have been made in different parts of the world to attempt to minimize it. This book presents some conceptual and practical approaches to respond to this phenomenon with a particular focus on the tropics.

The book is a compilation of selected articles presented in the 2015 Tanri Abeng University Conference in Jakarta, Indonesia. This international conference was organized jointly by the School of Architecture, Tanri Abeng University, Indonesia; London Metropolitan University; England, and Heriot-Watt University, Scotland.

Seventeen articles have been selected as chapters which are grouped into four parts to form the book. The first part of this book deals with the general issue of climate change. It discusses the causes and the ways of mitigation and adaptation of the built environment to climate change. This first part discusses the development of some developing countries, such as Indonesia, in which improving a nation's well-being would need a massive development of the nation's infrastructure and its built environment. The massive developments of infrastructure and the built environment have triggered the use of fossil fuel and other earth's resources, emitting huge amounts of carbon dioxide, creating global warming and climate change. Some conceptual strategies to overcome these problems are discussed in this part. The concept of sustainability, in which reducing carbon emissions was seen as likely to be the only way to reduce global warming, is challenged. The renewable energy sources, which emit low carbon, are still demanding land to generate the energy, raising the ecological footprint. There are articles in this part that discuss the way to mitigate climate change in non-tropical climates, and these can be used as a comparative study for a tropical case.

The second part of this book offers articles that examine the way to overcome disasters in the city caused by climate change. The Indonesian capital city of Jakarta has suffered from frequent floods, creating problems not only for the fishing settlements in the coastal areas but also for those living in the centre of the city. Some proposed strategies to overcome these problems are discussed in this part. The way to reduce the urban heat island (UHI) effect in the urban areas is also discussed in this chapter. It deals with the Singaporean problems of land being covered mainly by hard surface materials across the country. Along with the problems of flood, Jakarta is also suffering from traffic congestion, creating air pollution in the capital city. In this part, a bicycle lane system is proposed to improve the air quality and at the same time reduce the traffic congestion and the carbon emissions in this city.

The third part of the book deals with some practical approaches to make buildings that can help to mitigate climate change by reducing their need for nonrenewable energy. It proposes ways to reduce cooling energy by means of passive design in tropical buildings, such as in Malaysia. Examples are given of how buildings can be designed in such a way that solar heat gain is kept to a minimum, to create low indoor temperatures with a minimum help from mechanical means, thus reducing building energy consumption and promoting low-carbon buildings.

The last part of the book deals with the role of plants in mitigating and adapting the built environment to climate change. The impacts of plants as vertical and rooftop gardens to reduce the adjacent outdoor temperature as well as the indoor temperature in Singapore are discussed in this part along with the use of banana fibre as a possible source of building materials in Brazil.

I would like to thank all of the contributors who submitted their articles compiled in this book. Gratitude is given to Springer, the publisher which kindly publishes this book. I would also like to thank Fergus Nicol and Sue Roaf for their efforts to invite authors to submit their articles in our successful conference.

Jakarta, Indonesia Tri Harso Karyono

Contents

About the Editors

Professor Tri Harso Karyono
Tri was trained as an architect at the Department of Architecture, Bandung Institute of Technology (ITB), Indonesia, from which he received an engineering degree (Ir.) in architecture. He received a master's degree (MA) in architecture from the University of York, England, following the completion of his dissertation on "Solar Energy and Architecture: A Study of Passive Solar Design for Hospital Wards in Indonesia". Tri was awarded a doctorate (PhD) from the School of Architectural Studies, University of Sheffield, England, after finishing his thesis on "Thermal comfort and energy studies in multi-story office buildings in Jakarta, Indonesia".

Teaching in a number of architectural schools in Jakarta and also acting as doctoral external examiner at various universities, he has published a number of books on sustainable built environment and tropical architecture. His book on green architecture (in Bahasa Indonesia) (2010) has been reprinted three times within 6 years of publication. His latest book *Tropical Architecture: Form, Technology, Comfort and Energy Use* (in Bahasa Indonesia) (2016) tries to reposition the term of tropical architecture, which has been misinterpreted by many Indonesian scholars as a vernacular architecture. Tri has also been publishing a number of scientific articles, particularly related to thermal comfort in the warm and humid tropical climate in a number of international journals. He has published numerous articles in a number of Indonesian newspapers and architectural magazines. He designed his low-energy house at the outskirts of Jakarta in 2007. This house consumes very little electricity and is fairly comfortable without air conditioning.

Tri was awarded a professorship by the Indonesian Ministry of Education in 2007 and presented his inaugural lecture entitled "From Thermal Comfort to Global Warming: Architecture and Energy Points of View". His research interests are in the fields of thermal comfort, tropical architecture, sustainable built environment and low-carbon town.

Professor Robert Vale and Professor Brenda Vale

Robert Vale and Brenda Vale are architects and academics who studied architecture together at the University of Cambridge and wrote their first book on sustainable design, *The Autonomous House*, in 1975. Following their design of several award-winning buildings for the National Health Service in the UK, they built the UK's first autonomous house in 1993 and the first zero-emissions settlement, the Hockerton Housing Project, in 1998. It was the analysis of the performance of these buildings that revealed the importance of behaviour and led to their current research into ecological footprints, using the concept initially devised by Wackernagel and Rees in Canada in the 1990s.

The Vales have received a number of international awards, including those from the United Nations and the European Solar Energy Society, for their work. They carried out the initial development of the Australian government's National Australian Built Environment Rating System (NABERS) which has now been put into operation. Their 2009 book, *Time to Eat the Dog? The Real Guide to Sustainable Living*, used the ecological footprint to look at the environmental impact of how we live today, including the impact of household pets. Their surprising finding, subsequently supported by other researchers, was that a big dog in a western country has a similar impact to that of a person in Indonesia. More recently, they edited a book called *Living Within a Fair-Share Ecological Footprint* which comprised a series of studies written by colleagues, former students and current postgraduates. They are currently collaborating on a book that compares the ecological footprint of daily living around the world. Their areas of research at present are in the fields of sustainability, resilience, building materials and architectural history.

Chapter 1
Introduction: The Tropics: A Region Defined by Climate

Robert Vale and Brenda Vale

Abstract The Tropics is a region defined by geography rather than by political boundaries. This makes it a region also defined by climate. Global climate is changing as a result of greenhouse gas emissions from human activities. Temperature increases as a consequence of climate change are likely to be most severe in the Tropics, a region which is home to 40% of the global population, but produces only 15% of the world's greenhouse gas emissions. This book provides a wide range of examples that have been proposed by researchers to show the practical contributions that the Tropics can make in response to climate change. These contributions can assist in solving a problem largely imposed on the region by the rest of the world.

Keywords The tropics • Geography • Climate change • Effects of temperature

This book is about the impacts of climate change on the Tropics. The Tropics is a somewhat unusual area of the Earth's surface, because unlike most similar groupings of nations such as Europe or Australasia, it is a region defined entirely by physical rather than political boundaries. The Tropics as a zone is defined by two lines of latitude, the Tropic of Cancer in the north and the Tropic of Capricorn in the south. These imaginary lines are the two circles around the Earth that mark the latitudes of 23.5 degrees north, where the sun is directly overhead at noon on June 21st (midsummer in the northern hemisphere) and 23.5 degrees south where the noon

R. Vale (✉) • B. Vale
School of Architecture and Design, Victoria University of Wellington,
Wellington, New Zealand
e-mail: robert.vale@vuw.ac.nz

T.H. Karyono et al. (eds.), *Sustainable Building and Built Environments to Mitigate Climate Change in the Tropics*, DOI 10.1007/978-3-319-49601-6_1

sun is directly overhead on December 21st (midsummer in the southern hemisphere). The names of the two tropics are derived from the constellations in which the sun appeared at these times when the two lines were defined in the distant past (Jordan 2015).

Thanks to this physical definition the Tropics as a region comprises a much wider range of nations and cultures than most politically-derived groupings. The Tropics includes much of Central and South America, most of Africa, India, Pakistan and Bangladesh, Thailand, Cambodia, Vietnam and the southern tip of China, the whole of Indonesia and the Philippines and the top third of Australia. (Edelman et al. 2014; 7). It is a zone of great cultural differences and of great disparity of wealth and its occupants are numerous, the region is home to around 40% of the world population (Edelman et al. 2014; 15). What all these people share is not so much a common culture as a common climate.

This idea of a common climate for the Tropics is not entirely true although tropical regions are characterised by being hot, with the lowest monthly average temperature being 18 degrees C or higher. Nearly all areas in the Tropics fall into either the "tropical" or "dry" zones of the Köppen climate classifications (Chen and Chen 2013). However even in otherwise hot tropical areas there can be quite different local climates, for example potatoes are widely grown on the Dieng plateau in tropical Java, Indonesia, because it lies about 2000 metres above sea level and is consequently much cooler than the surrounding lower areas.

1 Limiting Climate Change by Limiting CO_2

The Tropics is already a region characterised by high temperatures, but with climate change, temperatures that are already high are likely to become even higher. The United Nations Environment Programme has stated the need to limit global average temperature rise to no more than 1.5 or at most 2 °C compared with the pre-industrial level to prevent serious social, economic and environmental problems by 2100 (UNEP 2010). According to the UK Meteorological Office, by the end of 2015 the global average temperature was already 1 degree Celsius above the pre-industrial figure (Met Office 2015). When measurements of carbon dioxide, the principal "greenhouse gas" causing this temperature increase, started to be made at the Mauna Loa observatory in Hawaii in March of 1958, the concentration of CO_2 in the atmosphere was around 315 parts per million (ppm) and it has risen every year since measurements began (US Dept. of Commerce et al. 2016). A thousand years ago, before the Industrial Revolution when carbon-based fossil fuels began to be consumed on a large scale, the level of CO_2 was around 280 ppm (Etheridge et al. 1996).

There is a clear link between the amount of CO_2 in the atmosphere and the average temperature, the more CO_2, the higher the temperature. Scientists and even politicians have been aware of this link for a long time. For example, a paper presented in 2005 refers to the European Council ruling back in 1996 that the increase in global average temperature should not exceed 2 °C. The paper goes on to say that a rise of even 2 degrees cannot be regarded as "safe" and the author concludes that stabilizing CO_2 in the atmosphere at 400 ppm carries a reasonable likelihood of staying below the 2 degree temperature limit (Meinshausen 2005). This is getting quite urgent as NASA reported in 2013 that the level of CO_2 in the atmosphere had reached 400 ppm (NASA 2013).

2 Limits Defined by the Paris Climate Agreement

Climate change is a problem on a world scale but while the politicians of the European Council 20 years ago seem at least to have grasped the need to halt the rise in global temperature, not all politicians have been so farsighted. Nearly 20 years after the European Council ruling, the CoP21, United Nations Framework Convention on Climate Change (UNFCC), 21st session of the Conference of Parties was held in December 2015 in Paris, France. After lengthy deliberations, politicians from 196 countries successfully negotiated the Paris Climate Agreement. This had the global goal of holding temperature increase to below 2° C above pre-industrial levels, with parties agreeing to make efforts to keep it below 1.5° C (Climate Action 2015). Meinshausen pointed out back in 2005 that a delay of even 5 years in implementing a reduction in emissions would make it much harder to avoid a harmful rise in temperature.

The Paris Climate Agreement is expressed in terms of global temperature rise while concentration of CO_2 in the atmosphere is usually expressed in parts per million (ppm), or more correctly as parts per million by volume (ppmv) and the relationship between atmospheric concentration and possible temperature increase is not precise. The Intergovernmental Panel on Climate Change (IPCC 2007) refers to

> …the likelihood of exceeding an equilibrium temperature threshold of 2 °C above pre-industrial levels based on a range of published probability distributions for climate sensitivity. To render eventual exceedance of this exemplary threshold 'unlikely' (<33% chance), the CO_2-equivalent stabilisation level must be below 410 ppm for the majority of considered climate sensitivity uncertainty distributions (range between 350 and 470 ppm).

The IPCC stated some time ago

> ... to achieve a 450ppmv concentration target, average carbon emissions per capita globally need to drop from about 1 tonne today to about 0.3 tonnes in 2100... (Banuri and Weyant 2001, 89)

Even to limit future atmospheric CO_2 to 450ppmv, 50ppmv higher than at present, average global emissions need to fall by 70%. However it should be noted that James Hansen of NASA and co-researchers made the point in 2008 that a limit of 450ppmv is too high and that a target of 350ppmv is needed to avoid severe problems in the future. (Hansen et al. 2008). To meet this target would mean zero carbon emissions (Baer et al. 2009).

As the UK Met Office has stated, with an average temperature now 1 degree C higher than normal, humanity is already two thirds of the way towards the UN's lower limit and even with this apparently small increase already signs of a changing climate can be seen. As the US Environmental Protection Agency states " Extreme weather events are becoming more frequent and/or severe around the world. This is consistent with what we expect with a warming planet" (EPA 2016). It is not only the United Nations that has shown concern about climate change, the World Bank has described climate change as "the biggest market failure in human history" (World Bank 2009, 61).

3 Climate Change in the Tropics

It appears highly likely that the effects of climate change will be more severe in the tropics than in the cooler parts of the world. Examples of this concern have appeared in the literature over an extended period. For example in 2009 Kjellstrom et al. (2009) suggested that temperature rise as a result of climate change in countries that are already hot could result in people being unable to work when exposed to such temperatures, which would threaten agriculture, construction and other outdoor activities. Reacting to comments by Pope Francis at the time of the Paris Conference that the poor would suffer most from climate change, Martin, writing in the *MIT Technology Review*, supported this view, listing drought, flooding and disease as the principal climate-related threats to human life and agriculture in the Tropics. The Tropics are more at risk because, as Martin concluded, "normal temperature ranges in the tropics fall within a narrower range than those in more northern climes, and so any deviation is likely to have more significant effects" (Martin 2015). This is supported by the work of Mora et al. at the University of Hawaii, who found that "..." Unprecedented climates "[ie. where climate exceeds the bounds of historical variability] will occur earliest in the tropics" (Mora et al. 2013). Their study found that the first city to experience this unprecedented climate, which they predicted to happen by 2020, would be Manokwari, the capital of West Papua, Indonesia, and before 2030 Singapore, Jakarta, Kuala Lumpur and several other major cities in the Tropics would be in the same situation (Moralab n.d.). It will not be long before many tropical cities will be experiencing climates unlike anything they have experienced before.

According to UN Habitat the built environment of cities is a source of much of what drives climate change; " Cities are major contributors to climate change: although they cover less than 2 % of the earth's surface, cities consume 78% of the world's energy and produce more than 60 of all carbon dioxide and significant amounts of other greenhouse gas emissions, mainly through energy generation, vehicles, industry, and biomass use." (UN Habitat 2012) The same source shows that the built environment is not only a contributor to climate disruption but also one of its victims; "cities and towns are heavily vulnerable to climate change. Hundreds of millions of people in urban areas across the world will be affected by rising sea levels, increased precipitation, inland floods, more frequent and stronger cyclones and storms, and periods of more extreme heat and cold." (UN Habitat 2012).

In spite of the fact that the Tropics is home to 40 % of the global population, as a region it produces only 15 % of the world's greenhouse gas emissions (Edelman et al. 2014; 15). It seems unfair that the region of the earth which will likely suffer the most as a result of climate change due to emissions that it has not caused is still required to play its part in reducing emissions. To show the contribution that the Tropics can make in solving a problem that is largely not of its own making, this book provides a wide range of examples that have been proposed by researchers.

References

Baer P, Athanasiou T (EcoEquity), Kartha S (Stockholm Environment Institute) (2009) A 350 ppm emergency pathway: a greenhouse development rights brief Stockholm, Eco Equity/Stockholm Environment Institute

Banuri T, Weyant J (2001) Setting the stage: climate change and sustainable development in working group III: Mitigation Geneva, IPCC p89

Chen D, Chen H (2013) Using the Köppen classification to quantify climate variation and change: an example for 1901–2010. Environ Dev 6(1):69–79

Climate Action (2015) Find out more about CoP21. Available at http://www.cop21paris.org/about/cop21. Accessed Feb 2016

Edelman A, Gelding A, Konovalov E, McComiskie R, Penny A, Roberts N, Templeman S, Trewin D, Ziembicki M, Trewin B, Cortlet R, Hemingway J, Isaac J, Turton S (2014) State of the tropics 2014 report. Report. Cairns, James Cook University

EPA (2016) Understanding the link between climate change and extreme weather. Washington, DC, United States Environmental Protection Agency. Available at https://www3.epa.gov/climatechange/science/extreme-weather.html. Viewed 26 May, 2016

Etheridge DM, Steele LP, Langenfelds RL, Francey RJ, Barnola J-M, Morgan VI (1996) Natural and anthropogenic changes in atmospheric CO_2 over the last 1000 years from air in Antarctic ice and firn. J GeophysRes 101(D2):4115–4128

Hansen J, Sato M, Kharecha P, Beerling D, Berner R, Masson-Delmotte V, Pagani M, Raymo M, Royer DL, Zachos JC (2008) Target atmospheric CO_2: where should humanity aim? Open Atmos Sci J 2:217–231. doi:10.2174/1874282300802010217

IPCC Climate Change (2007) Mitigation. Contribution of working group III to the fourth report of the intergovernmental panel on climate change. Cambridge University Press, New York

Jordan C (2015) What is the significance of the Tropic of Cancer, Tropic of Capricorn, Arctic Circle and Antarctic Circle?. In Ask an astronomer Cornell University, June 27. Available at http://curious.astro.cornell.edu/physics/160-our-solar-system/the-earth/seasons/15-what-is-the-significance-of-the-tropic-of-cancer-tropic-of-capricorn-arctic-circle-and-antarctic-circle-beginner. Viewed 16 June 2016

Kjellstrom T, Holmer I, Lemke B (2009) Workplace heat stress, health and productivity – An increasing challenge for low and middle-income countries during climate change. Global Health Action, [S.l.], V. 2, Nov 2009. Available at: http://www.globalhealthaction.net/index.php/gha/article/view/2047. Viewed 25 May 2016

Martin R (2015) Climate change: why the tropical poor will suffer the most. MIT Technology Review: Energy. 17 June Available at https://www.technologyreview.com/s/538586/climate-change-why-the-tropical-poor-will-suffer-most/. Viewed 26 May 2016

Meinshausen M (2005) On the risk of overshooting 2 °C. Paper presented at scientific symposium "Avoiding dangerous climate change", Exeter, Met Office, 1–3 February. Available at http://

www.pik-potsdam.de/~mmalte/simcap/publications/meinshausenm_risk_of_overshooting_final_webversion.pdf. Viewed 16 June 2016

Met Office (2015) Global temperatures set to reach 1 °C marker for first time. 9 Nov. Exeter, Met Office. Available at http://www.metoffice.gov.uk/news/release/archive/2015/one-degree. Viewed 26 May 2016

Mora C, Frazier A, Longman R, Dacks R, Walton M, Tong E, Sanchez J, Kaiser L, Stender Y, Anderson J, Ambrosino C, Fernandez-Silva I, Giuseffi L Giambelluca T (2013) The projected timing of climate departure from recent variability. Nature 502:183–187. 10 October

Moralab (n.d.) The year of climate departure for world cities. University of Hawaii at Manoa, Honolulu. Available at http://www.soc.hawaii.edu/mora/PublicationsCopyRighted/Cities%20Timing.html. Viewed 26 May 2016

NASA (2013) NASA scientists react to 400 ppm carbon milestone. 21 May. Pasadena, CA, Earth Science Communications Team at NASA's Jet Propulsion Laboratory, California Institute of Technology. Available at http://climate.nasa.gov/400ppmquotes/Viewed. 20 June 2016

U.S. Department of Commerce, National Oceanic and Atmospheric Administration Earth System Research Laboratory, Global Monitoring Division (2016) Full Mauna Loa CO_2 record. Available at http://www.esrl.noaa.gov/gmd/ccgg/trends/full.html. Viewed 15 June 2016

UN Habitat (2012) Urban themes: climate change. United Nations Human Settlements Programme, Nairobi. Available at http://unhabitat.org/urban-themes/climate-change/. Viewed 26 May 2016

UNEP (2010) The emissions gap report; Are the Copenhagen accord pledges sufficient to limit global warming to 2 °C or 1.5 °C?. United Nations Environment Programme, Kenya. Available at www.unep.org/publications/ebooks/emissionsgapreport

World Bank (2009) World development report 2010: development and climate change. The World Bank, Washington, DC

Part I
The Sustainability of the Built Environment

Chapter 2
Climate Change and the Sustainability of the Built Environment in the Humid Tropic of Indonesia

Tri Harso Karyono

Abstract As a developing country, Indonesia needs to develop its facilities and infrastructure to improve its people's wellbeing. As a consequence of the development, much of its energy sources, particularly its fossil fuels, have to be consumed, emitting a huge amount of carbon dioxide, accelerating global warming and climate change. In many cases, development would increase a country's carbon emissions and its ecological footprint (EF), which could threaten the sustainability of the whole earth. Indonesia is no exception; the development of this nation would emit much more carbon dioxide as around 95 % of this country's energy demand is supplied currently by fossil fuels. Considering the fact that Indonesia has a huge potential of renewable energy sources, substituting the fossil fuels by renewable sources would help this country to develop without emitting so much carbon dioxide, thus retarding the effects of climate change. Along with the diversification from fossil fuels to renewable energy, changing the human attitudes so as to behave wisely to use less energy would help this country keep its EF low, thus sustaining its development without adding to global warming and climate change.

Keywords Carbon emission • Climate change • Development • Fossil fuels • Renewable energy

1 Introduction

Many disasters have occurred in many places all over the world within the last decades. This gives a strong hint that something has changed on this earth. The impact of global warming is clear and the global climate has already changed. The changes of global climate give many consequences to the earth and thus for people's lives. Heavy rains create big floods which now occur not at a common time and not in common places. Forests burn not in the dry season as happened before, fire occurs more frequently at any time even during the rainy season. Anything can occur

---

T.H. Karyono (✉)
School of Architecture, Tanri Abeng University, Jakarta, Indonesia
e-mail: t_karyono@yahoo.com

T.H. Karyono et al. (eds.), *Sustainable Building and Built Environments to Mitigate Climate Change in the Tropics*, DOI 10.1007/978-3-319-49601-6_2

anywhere at any time. Weather forecasters may become unable to forecast as unexpected factors may change everything unpredictably.

Looking at the built environment aspect, the cause of climate change is obvious. In addition to the increase of human population, so many new things and activities have been created since the industrial revolution. The addition of human population and the new activities created by human beings require new infrastructures, facilities and buildings to be built to accommodate them. Most of these things are designed or proposed by architects. In a simple view, the work of architects is to craft the earth's surfaces. Unlike the ordinary craftsmen, architects have to craft them with huge responsibilities laid on their shoulders. Architects have to create these kinds of products such as buildings, bridges, towns, etc., to be aesthetic, but also to be functionally correct, and at the same time, these things have to minimize the use of earth's resources, to consume less energy and emit less carbon. Strictly speaking, architects have to consider that crafting the skin's earth must be ecologically friendly, conserving earth's resources, minimizing the use of energy, minimizing carbon emissions, producing less waste, and last but not least, not sacrificing the quality of human life to have a safe, healthy and comfortable living environment.

In many cases, architects fail to do this. Many of them have built buildings, towns and other man-made environments which have degraded the environment in many ways. They unconsciously destroy the natural environments, whether seaside, swamps, lakes, rivers or forests, by converting them into buildings, housing, offices, commercial facilities, and even towns. Figure 2.1 shows a reclamation area on the

Fig. 2.1 A reclamation area of Pantai Indah Kapuk, Jakarta: Architects and planners destroy a natural coastal environment, creating urban heat islands and raising the ambient temperature of the city

seasides of North Jakarta. The developer and the local authority do not consider this development may have degraded the natural environment and the ecological system. Many green areas, with an abundance of trees and wildlife, have been removed, replaced with concrete, asphalt and other hard surface materials. This has created changes in the micro climate in many parts of the earth's surface. Ambient temperatures rise, clean water runs out and the wildlife disappears. The changes of the earth's skin environment from natural to man-made have created huge environmental problems in many ways. The urban heat island phenomenon, in which the built up areas are becoming much warmer than their surrounding environments, has taken place everywhere.

On the other hand, the increased number of the human population and the rapid increase in use of modern technology has increased the energy demand. Most of all current human activities require energy, which is quite unfortunate given that the energy being used today is mostly derived from fossil fuels, emitting a huge amount of carbon dioxide in the generating process. The huge emission of CO_2 is scientifically proven to be the cause of global warming and climate change. The latest data of The World Bank in 2013 showed that around 81.2 % of the world's energy use is generated from fossil fuel (World Bank 2016), the rest from alternative, renewable and nuclear energy.

All the combined problems have to be learned and to be solved. How could we build buildings and other man-made environments, such as a town, that can be safe, healthy and comfortable, without resulting in the degradation of the environment, emitting less carbon dioxide and preventing the rise of keeping the earth's temperature?

The recent development of Indonesia, as in other developing countries, has concerned environmentalists that the development has degraded the country's natural environments. Development has produced many new buildings, settlements and towns. However, unlike in the past, in which our ancestors had been successfully handling the environment wisely by limiting the use of natural resources, the new development has encouraged people to use more energy and natural resources. The trained Indonesian architects and urban planners are unlikely to do as wisely as our ancestors did. Today, many buildings and towns are designed without considering the local climate and culture. Furthermore, the lack of reference to informal wisdom and the poor formal building codes have created environmental problems in many ways, such as floods, traffic congestion and so on. The worst part of the development of Indonesia is that it has encouraged people to use more energy, emitting more CO_2, creating global warming and climate change. Some problems caused by climate change, such as floods, have afflicted many large Indonesian cities, such as Jakarta. In order to mitigate climate change, some strategies are proposed below.

2 Indonesia as a Tropical Developing Country

Indonesia lies between the Pacific and Indian Oceans, it stretches along the equator and scattered over both sides for a distance of 6400 km, and lies between 11° south latitude and 6° north latitude, and 95° and 141° east longitudes (Wikipedia, Indonesia 2016a). The largest archipelago in the world, Indonesia is formed of 17,508 islands, only about 6000 of which are inhabited (CIA 2016). Java, the fifth largest island in Indonesia after Kalimantan, Papua, Sumatera and Sulawesi, is the most densely populated island in Indonesia. More than 135 million out of the country's 240 million population live on this island. Java is recorded as the world's most populous island (Wikipedia, Java 2016c). Figure 2.2 shows the Mount Kie Matubu in Tidore Island, North Maluku, Indonesia. Hundreds of mountains and volcanoes in Indonesia are not only shaping the beauty of its landscapes, but also fertilize the soil, creating a perfect closed chain of the ecological system.

With around 127 active volcanoes (Wikipedia, list of volcanoes 2016f), Indonesia is the country with the largest number of active volcanoes in the world. With so many volcanic eruptions over a long time, the volcanoes' ashes help to fertilize the country's soils. People can easily grow vegetables, plants and trees anywhere in this country without fertilizer. There is a big potential for the Indonesians to live autonomously without depending on the global food industry.

With 98 million hectares of rainforest (Orangutan Foundation 2016), 326 million hectares of sea and 9.3 million hectares of freshwater (Wikipedia, Geography 2016e), this country could absorb huge amounts of CO_2 annually. The rainforest could take up 5 tonnes of CO_2 per hectare per year (Bao and Phan 2011), which means all the Indonesian forest could help to reduce CO_2 annually by 490 million tonnes.

Fig. 2.2 Mount Kie Matubu in Tidore Island, at the back of the small island of Maitara, North Maluku, Indonesia

The forests of Indonesia represent 10 % of the world's remaining tropical forests. Indonesia's forests are the second largest in the world after Brazil (Orangutan Foundation 2016). The diversity of Indonesia's flora and fauna and the fertility of its soil create a perfect closed chain of the ecological system. Figure 2.3 shows Maribaya forest, West Java, Indonesia. When the Indonesian natural environment, forests and the sea are conserved, they will curb a quite large proportion of world carbon emissions.

However, the country's potentials have been neglected somehow. The country's recent economic development has increased people's ability to get more things. Along with this, the modern economy has driven people to buy things that they do not actually need. Modern food technology has offered fancy products that attract people to consume them. Within a few decades, there has been a changing of Indonesians behaviour towards life, from practicing a simple and traditional way of life to a modern life. The changes of attitude towards life change the use of technology to support that life. Modern technology is an avoidable thing to be used as it helps people to have everything faster and easier in some way.

Modernity enhances people's ability to increase their living standards, increasing the quantity of earth's resources to be consumed. Forests are cut down to allow the development of massive plantations, such as palm oil, and also for new industry.

Fig. 2.3 Maribaya forest, West Java, Indonesia. The tropical rain forest is a perfect closed chain of the ecological system

On the other hand, huge areas of paddy-rice fields have been replaced by houses, buildings and even new towns. Indonesia has started to be a food importing country as the food production has declined while its demand has increased.

3 Problems of Improving the Wellbeing in the Developing Countries

Like most of the developing countries, Indonesia is no exception. This country needs to improve its people's wellbeing to be more prosperous. To achieve this, this country needs to build many more infrastructures, buildings and other facilities. More roads, railways, bridges, are needed to connect towns and places. More houses need to be built to accommodate people and population growth. More schools are needed to be built, as well as hospitals, public libraries, and other public facilities to facilitate the society's needs. Figure 2.4 shows Muara Angke settlement with a background of exclusive apartments. The gap between the poor and the rich is quite wide, while the poor want to have an equal opportunity.

On the other hand, the prosperity of the nation is likely to be associated with a country's gross domestic product (GDP) and human development index (HDI). Prosperous countries tend to have GDP and HDI higher than those of the poor countries. The developed industrialized countries tend to have GDP and HDI higher than the developing countries. Improving a nation's wellbeing for a developing country such as Indonesia, means that many things have to be built and developed, which in turn requires a lot of energy and natural resources. Increasing the use of energy would increase the country's carbon dioxide (CO_2) emissions since most of the

Fig. 2.4 Muara Angke settlement with a background of exclusive apartments, Jakarta

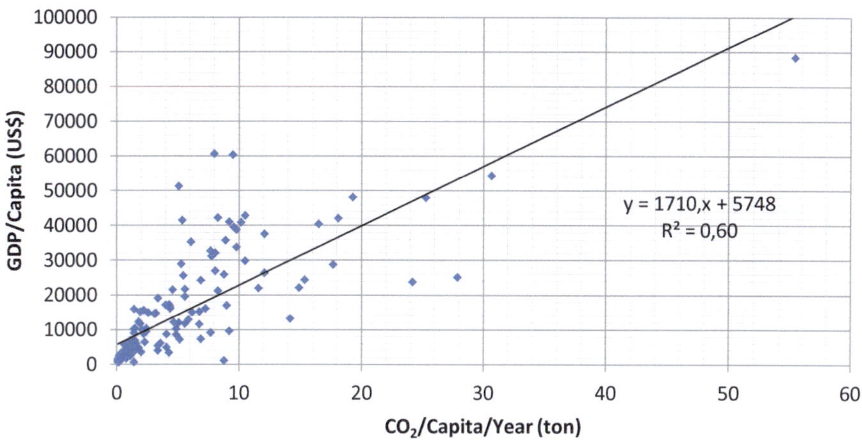

Fig. 2.5 Linear regressions of GDP on CO_2 (Revision of Karyono's chart 2013)

energy in Indonesia is derived from fossil fuels. At the same time, increasing the use of natural resources would increase the country's ecological footprint (EF).

There is clear evidence that improving a nation's well-being, in terms of the gross domestic product (GDP) and human development index (HDI), would increase the nation CO_2 emissions and EF. Examining 143 countries with the available data on their GDP (Wikipedia, GDP 2013, Countries of the World 2009), HDI (Wikipedia, HDI 2016b), CO_2 (Wikipedia, CO_2 2016d) and (EF) (Global Footprint 2010), it was found that the above indexes are correlated with each other. The calculation is based on the data of countries' GDP, HDI, CO_2 and EF in 2007. This date has been used since the world's EF is most recently reported for this year. To make all the data comparable, the analysis must be using the same time, that is 2007. Figure 2.5 shows a graph of the correlation between GDP (US$) and$CO_2$ emissions (tonnes/capita) taken from 143 countries with the available data. The regression equation is expressed as:

$$GDP = 1710\ CO_2 + 5748 \tag{2.1}$$

With the coefficient of determination (R^2) of the line being 0.6, it shows that there is a strong correlation between GDP and CO_2 emissions. Although each country has its own specific correlation between its GDP and CO_2 emissions as it is based on the energy sources being used by the country, an increase of a nation's GDP is likely to increase the country's CO_2 emissions. France is an exception. This country emits a relatively low amount of CO_2 to increase its GDP. Around 78 % of France's electricity is generated by nuclear power, emitting low CO_2 emissions. Having a GDP of US$ 35,247 per capita, France emits only 6.1 tonnes CO_2 per capita, which is about 0.2 kg CO_2 for each single dollar to raise its GDP. This is the lowest CO_2 production per unit GDP among the world's major industrial countries (Wikipedia, CO_2 2016d). In order to keep a low CO_2 emission value a country must minimise the

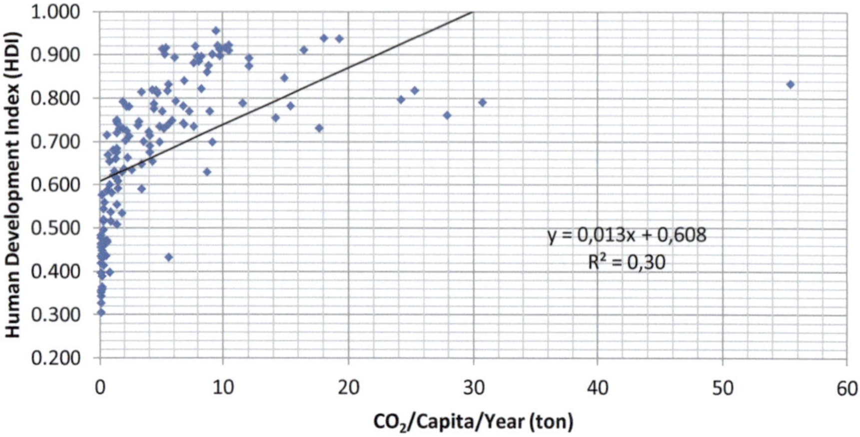

Fig. 2.6 Linear regressions of HDI on CO_2 (Revision of Karyono's chart 2013)

use of fossil fuels. Diversifying from fossil fuels to nuclear power as France does may not be recommended for Indonesia as this country has a huge potential of renewable energy sources, which can be developed to substitute its fossil fuel consumption.

A Pakistani economist, Mahbub ul Haq, developed a Human Development Index (HDI) in 1990, which is used by the United Nations to partly express the wellbeing of any country in the world. To see the relationship between a country's HDI and its CO_2 emissions, a regression line of HDI on CO_2 emissions is constructed. Figure 2.6 shows a linear regression line of HDI on CO_2. The regression equation is expressed as:

$$HDI = 0.013\ CO_2 + 0.597 \tag{2.2}$$

The coefficient of determination of the regression (R^2) is 0.3, or 0.55 in terms of the coefficient of correlation. The correlation of HDI and CO_2 is less strong than that between GDP and CO_2. The higher the HDI of a nation the more the nation would tend to emit CO_2.

William Rees and Mathis Wackernagel of the University of British Columbia, Canada, introduced the terms of Ecological Footprint (EF) in 1992 to measure humanity's earth's resources consumption. EF, which is measured in global hectares (gha), is a calculation of human demand on the earth's resources, standardized by land and sea areas that can regenerate resources being consumed by the population and mitigate associated waste (Global Footprint 2010). The higher the EF, the more the individual or nation consumes the earth's resources and produces more waste. Using EF as a parameter to measure national demands for the earth's resources, GDP and HDI seem closely related to the nations' EF levels. A nation with high GDP and HDI tends to have a high EF, which means that GDP and HDI tend to have a positive correlation with the consumption of earth's resources.

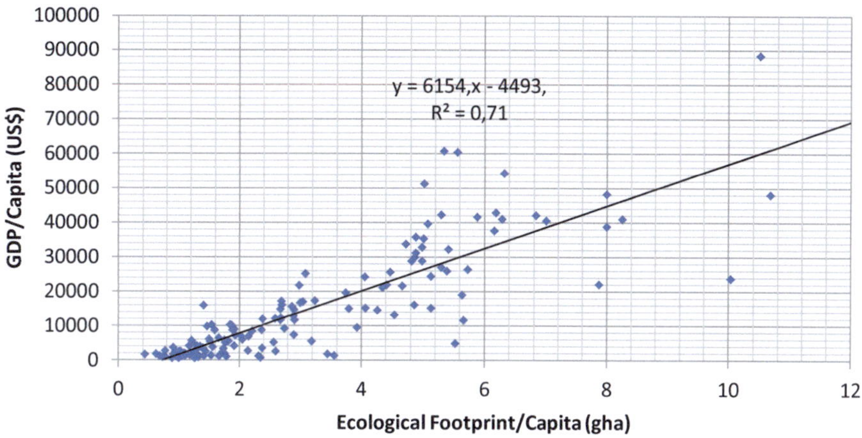

Fig. 2.7 Linear regression of GDP on EF (Revision of Karyono's chart, 2013)

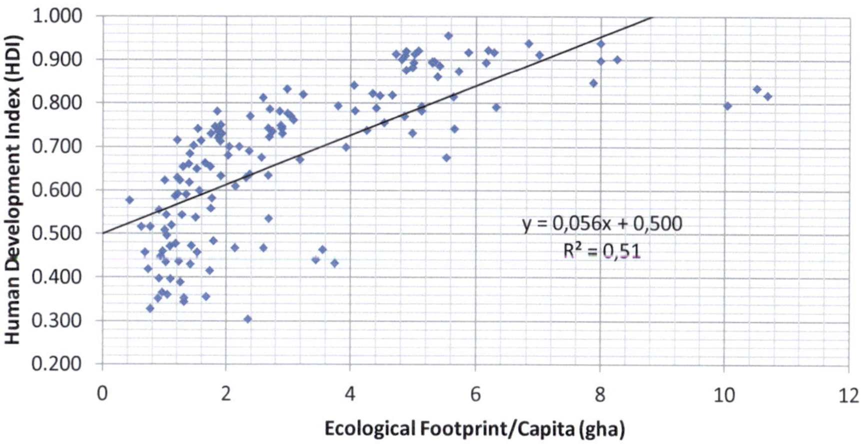

Fig. 2.8 Linear regression of HDI on EF (Revision of Karyono's chart, 2013)

In terms of GDP, Fig. 2.7 shows a linear regression line of GDP on EF. The regression equation is expressed as:

$$GDP = 6168EF - 4602 \tag{2.3}$$

Having 0.7 as a coefficient of determination (R^2) or 0.85 as a coefficient of correlation (r), the correlation between GDP and EF is quite strong. The higher the GDP of a nation, the higher the demand for earth's resources, requiring more land to be used. The correlation between GDP and EF is even stronger than the correlation between GDP and CO_2 emissions.

In terms of HDI, Fig. 2.8 shows a regression line of HDI on EF. The regression equation is expressed as:

$$HDI = 0.058EF + 0.487 \tag{2.4}$$

Having a coefficient of determination of 0.5 or coefficient of correlation of 0.7, the higher the HDI the higher the nation would consume the earth's resources. To improve a nation's well-being, there would be an increase in consumption of earth's resources, requiring more area of land to be used and increasing EF levels.

Generating electricity with nuclear power and renewable energy, such as wind turbines, solar PV or hydro would help the country in reducing its CO_2 emissions. However, it might not help too much with the EF reduction, since any energy generator would occupy a certain amount of land. The more energy is generated the more area of land is needed, increasing EF levels. In order to reduce the EF, a country must reduce its resource consumption, including food and minerals as well as energy. Reducing a nation's energy consumption, instead of diversifying to renewable and nuclear energy, is perhaps the best way to reduce a country's EF.

4 Climate, Development and Energy Consumption

To mitigate climate change, the world carbon emissions must be reduced. In the current situation, reducing carbon emissions would reduce the nations' GDP and HDI, degrading people's living standards and their wellbeing. The question is how to find a way of cutting a country's carbon emissions without degrading living standards? Improving the quality of the built environment is to create a healthier and comfortable environment. The improvement of our built environment could be done, even without the escalation of CO_2 emissions.

Climate affects people's life, behavior and in the end, their energy consumption. In a cool climate, people have to use more energy (electricity) for heating spaces, hot water, etc.,as without it, life would be miserable. In the winter, people have to encounter the low outdoor temperatures, making it very hard to live and work in the outdoor environment. Buildings are an important means to protect people from the unwanted cold weather. A building is not only a matter of a spatial form to accommodate people with their activities, but, a building is also a device which is used to modify an extreme outdoor environment to be moderate and comfortable enough to accommodate people keeping their activities at a normal metabolic rate.

To create a comfortable building, while the outdoor temperature is extremely low; heating up the indoor spaces is required. In this case, the use of energy for space heating is a must, making people highly dependent on the energy. This kind of circumstance, however, does not apply to a tropical climate, such as Indonesia. There is no need for heating up the rooms, though there might be an opposite action which must be taken, in which, cooling is necessary in some particular conditions.

Almost all of the regions of Indonesia tend to be warm, except those in the mountainous or highland areas where the air temperature is relatively low. Only a small number of Indonesians live in these cool areas. Thick clothes and space heating may be required by people in these areas.

On average, the prevailing moderate climate in the tropics, which is neither too cool nor too hot, has actually provided a greater chance for Indonesians to consume

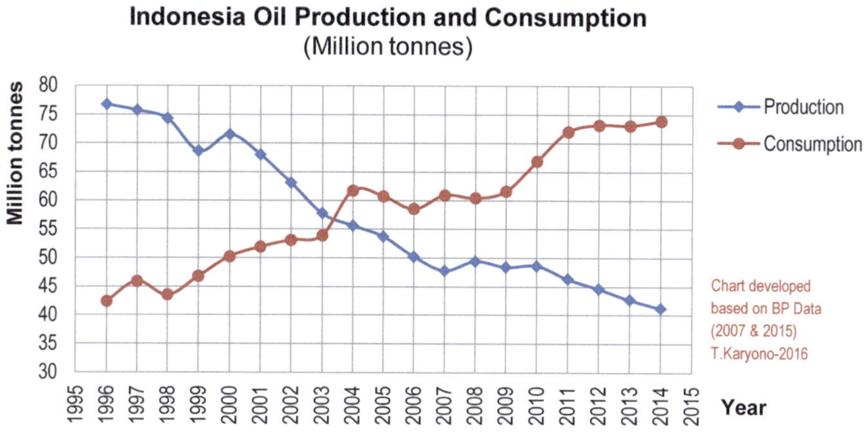

Fig. 2.9 Indonesia oil consumption surpassed its production in 2004

less energy than people who live in the temperate climate. To provide comfortable buildings, Indonesians need less energy than people who live in a country with more extreme climate condition, such as in the desert and near the poles. Having a prevailing outdoor temperature of between 18 to 32°C, which is relatively close to the skin temperature, many Indonesians living in the towns are relatively comfortable without heating or cooling. A number of homeless people can still sleep well in the open air on the roadsides with a minimum use of clothes. This would be certainly impossible for those living in the cold climate. A tropical building may need only a roof, no walls, to protect the occupants from rain and strong solar radiation. Since the dependency on energy is small, people in Indonesia have a better chance to sustain their built environment since they could burn less fossil fuel to provide a comfortable house and building. However, the facts have shown differently.

Along with the development of the nation, Indonesia's energy consumption has increased considerably since the New Order government took power in the middle 1960s. Along with the country's massive development since 1970, the country's oil consumption has risen steadily, while its production has continuously declined. In 2004, the oil consumption surpassed the production. The gap between them is getting bigger and bigger since Indonesia very much depends on oil (BP 2007, 2015). On the other hand, it is still fortunate for Indonesia that until now its natural gas and coal productions are still over the consumption. The new data show that in 2014 the country's natural gas production was 66.1 million tonnes oil equivalent (toe) and the consumption was 34.5 million toe, while the coal production was 281 million toe and the consumption was 60.8 million toe (BP 2015). Figure 2.9 shows Indonesia oil consumption surpassed its production in 2004.

Apart from producing a proportion of fossil fuels (oil, gas and coal), this country has abundant of renewable energy sources, such as geothermal, solar, hydro and biomass. However, this huge potential of renewable energy sources is still far from being utilised. The country's electricity still largely depends on fossil fuel. Nearly

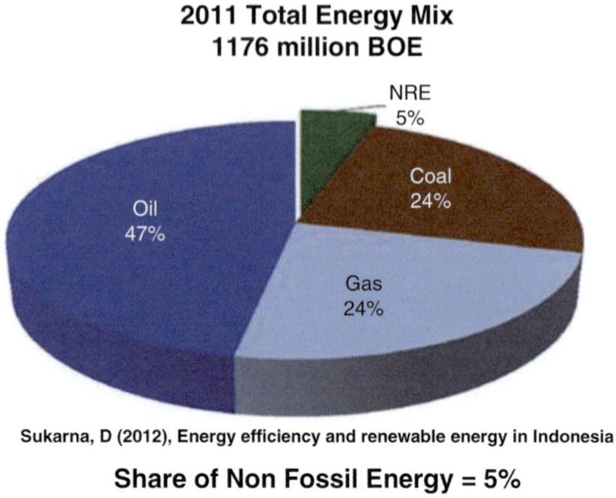

Fig. 2.10 A large proportion of non-renewable energy is still used (Sukarna 2012)

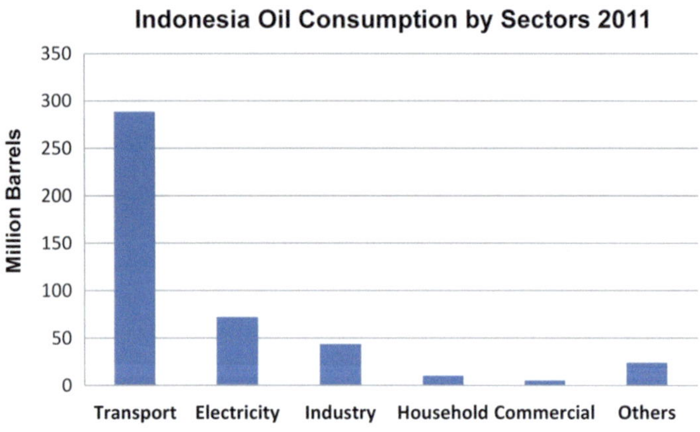

Fig. 2.11 Indonesia oil consumption by sectors in 2011 (Chart developed based on Indonesia Ministry of Energy data 2012)

all of the energy used in the built environment comes from fossil fuels. Figure 2.10 shows a small proportion of renewable energy is generated to supply the nation's energy demand. Only about five percent of the total national energy supply comes from renewable sources, the rest 95% from fossil fuel, in which 47 % is oil, 24 % gas and 24 % coal (Sukarna 2012).

Figure 2.10 shows oil covers the biggest portion (47 %) of energy sources used in Indonesia in 2011, while Fig. 2.11 shows Indonesia's oil consumption by sectors in 2011. It can be seen from this Table that transportation consumed the highest portion of the country's oil consumption (65 %), followed by electricity (16 %), industry (10 %), household (2 %), commercial (1 %) and others (6 %), which includes agriculture, mining, etc. (Indonesia Ministry of Energy 2012).

Transportation has dominated the country's oil consumption as most of the people tend to use private motorized vehicles, rather than public transport, bicycle or foot. The possible reasons for this, firstly, fuel is affordable for most people who have motorized vehicles, secondly, public transport in most of the Indonesian towns is not convenient, and lastly there some weaknesses in designing Indonesian cities and towns across the country. Cities are designed to give more priority to people with private cars. Every building must have a minimum car space as a mandatory requirement from the city authority. This has encouraged people to use private cars as they are plenty of spaces in the buildings for a parking lot. Within three decades, between 1987 and 2010 the number of motorized vehicles in Indonesia has escalated dramatically and is now around ten to twelve times greater, from 7,981,480 in 1987 to 76,907,127 vehicles in 2010 (Indonesia Statistical Agency 2012). On the other hand, cities have no adequate sidewalks for pedestrians and no bicycle tracks on the roads, making people reluctant to walk and cycle. Private cars and motorbikes dominate the roads, creating heavy congestion, consuming more fuels in transport.

In terms of the electricity sector, oil is consumed in the generation of the whole country's electricity, which is run by the government electricity company. Oil is still the main energy source for the country'selectricity. Oil consumption for the industrial sector is mainly used for the electricity generation of the local premises, such as factory, storage, and office buildings, which have no connection with the state electricity grid. In the household sector, kerosene is the main oil product that is largely used for cooking and lighting, particularly in the remote villages, which are not connected to the state electricity grid. Many villages in the eastern part of Indonesia have a limited access to electricity. In 2007 the Indonesian government introduced LPG (Liquified Petroleum Gas) to substitute for kerosene for cooking, which then slightly reduced the use of kerosene, but not the entire fossil fuel consumption (Indonesia Ministry of Energy 2012). In the commercial sector, automotive diesel oil (ADO) is mainly used as the energy source for electricity generation in the big buildings, such as shopping malls, offices and storage.

5 Mitigating Climate Change Without Sacrificing the Country's Development

In the architectural world, climate change has been anticipated with the green movement. Many building owners demand to have greener buildings in order to reduce carbon emissions and minimise other resource consumption. Reducing carbon emissions can be done by minimizing the use of energy in buildings and changing the source of energy from fossil fuel to renewable.

Reducing energy consumption is sometime associated with the degradation of life quality, as more activities might be carried out by human muscle power. However, if people accepted the idea that a high quality of life does not necessarily mean reducing the use of muscle power, people may be agreed that reducing energy may affect not too much their life quality. At a building scale, providing a mechanical means like an elevator, escalator, paternoster, etc., to move people up and down would undoubtedly make the building's users more comfortable, however, it does not equally mean it is improving the quality of life if the building is only a few storeys. Taking the stairs, instead of an elevator, to move up and down in a four-storey building would minimize building energy consumption without degrading the quality of life and the exercise would be beneficial to people.

At the town scale, any town not only requires streets to make it healthy and comfortable, but also requires pedestrian pathways and bicycle paths for those wanting to move without motorized vehicles. When people can be easily and safely walking and cycling in the town using their own muscles, this would help the town to consume less energy without degrading people's life quality. However, in current Indonesian urban culture, using the stairs instead of the elevator, taking a walk instead of driving a car are seen to be as life degradation. In this sense, improving life quality is something equal to reducing the use of muscle power. This kind of value must be reversed and changed if we want to reduce the carbon emission without stopping the development process to improve the living standard for those who need it.

In most of the Indonesian regions the tropical climate provides a moderate ambient temperature. The ambient temperatures are relatively close to the human comfort requirement. Having an appropriate design, a building could achieve indoor thermal comfort by natural means, without AC. A number of Indonesia towns are located in the highlands with relatively low ambient temperature. Based on Karyono's comfort study in Bandung, with the range of the outdoor temperature of between 18 °C and 29 °C, respondents were comfortable at about 24.7 °C, which is close to the daily average temperature (Karyono 2008). Theoretically, people in this climate would require no energy for either heating or cooling. With proper design, providing a comfortable and convenient building and its environment can be achieved without consuming much energy.

Looking back at the nature of the indigenous Indonesian people's life, they are likely to live traditionally and respecting nature. They minimise the use of technologies operated by non-muscle power. People in Kampung Naga are still living modestly, using limited devices to carry out their life. The Inner Baduy are even more extreme to reject modern technology in their life. They are prohibited to use any means of transportation, including motorized and non-motorized vehicles, bicycles, horses and other animals. Using electricity and fossil fuel is prohibited. They have to walk for hours and even days to get to destinations.

Considering the Baduy's way of life towards nature may help Indonesia to retard climate change and sustain its built environments. Using muscle power to carry out human activities is not something backward. If we believe with this kind of value in our life, the use of energy can be reduced easily without it being felt that the quality of our life is degraded. Figure 2.12 Baduy people in the Outer Baduy village. People live modestly without the modern technology.

Fig. 2.12 Baduy people in the Outer Baduy village, Banten, Indonesia

Table 2.1 Indonesia potential renewable energy (Sukarna 2012)

Renewable energy	Potential resources (MW)	Installed capacity (MW)	Ratio (Installed/Resources) (%)
Hydro	75,670	6,654,29	8.8
Geothermal	29,038	1,226	4.2
Micro hydro	769,69	228,963	29.75
Biomass	49,810	1618	3.25
Solar energy	4,80 kWh/m^2/day	2245	–
Wind	3–6 m/s	187	–

Of course, in some cases, some developments absolutely require energy, apart from human muscle power. This country still needs a lot of roads and bridges to connect places in the remote areas. They are many more buildings, houses and infrastructure to be built, in which heavy equipment should be used. In this case the use of energy to operate the equipment is avoidable. However, in order to reduce the carbon emissions in generating the energy, renewable energy should be taken as a priority. Therefore, other than reducing the use of energy, diversifying the energy sources in generating energy should be highly considered. Replacing fossil fuels by renewable energy is a must, so that the development can still be carried out without degrading the environment.

In fact, Indonesia has abundant sources of renewable energy, however, this country is still under-utilizing these sources. Except for micro hydro, any other sources have only been used in a very small proportion out of its resource potential (Sukarna 2012). Table 2.1 shows a small portion of energy in Indonesia is generated from various renewable energy sources, such as hydro, geothermal, solar, wind and biomass.

Table 2.2 Lifecycle greenhouse gas emission estimates for electricity generators, according to Benjamin K. Sovacool's comparison (Benjamin and Sovacool 2008)

Technology	Description	Estimate (g CO_2/kWh_e)
Wind	2.5 MW offshore	9
Hydroelectric	3.1 MW reservoir	10
Wind	1.5 MW onshore	10
Biogas	Anaerobic digestion	11
Hydroelectric	300 kW run-of-river	13
Solar thermal	80 MW parabolic trough	13
Biomass	Various	14–35
Solar PV	Polycrystalline silicon	32
Geothermal	80 MW hot dry rock	38
Nuclear	Various reactor types	66
Natural gas	Various combined cycle turbines	443
Fuel cell	Hydrogen from gas reforming	664
Diesel	Various generator and turbine types	778
Heavy oil	Various generator and turbine types	778
Coal	Various generator types with scrubbing	960
Coal	Various generator types without scrubbing	1050

Table 2.2 shows the significant reduction in emissions that comes from the diversification of electricity generation from fossil fuels to renewable energy. Renewable energy (Wind, Hydro, Biogas, Biomass, Solar PV and Geothermal) emits low CO_2 compared to fossil fuel (Coal, Diesel, Oil, Natural gas). The renewable energy sources emit only one fifteenth to one hundredth of the amount of CO_2 to generate the same amount of electricity. Diversification of energy (electricity) generation from fossil fuels to renewable energy would significantly reduce the country's carbon emissions without sacrificing the country's development process.

Given that Indonesia has a huge potential of renewable energy sources, which can be used as a substitute for its fossil fuels there is a path to mitigate climate change in the development of the country, as implementing this substitution would result in the emission of a much smaller amount of CO_2. However, to sustain this country in the development process, substituting fossil fuels with renewable energy sources would not be enough, as widespread use of renewable energy sources would require land to generate the sources, thus increasing the country's EF levels. Changing people's attitude and behaviour towards low energy activities would be a way to help Indonesia to sustain its development as well.

References

Bao Huy B, Pham Tuan Anh (2011) Forecasting CO_2 sequestration on natural broad-leaved evergreen forests in Daknong province, Tay Nguyen University, Daklak, Vietnam. http://www.socialforestry.org.vn/Document/DocumentEn/Article%20on%20CO2%20sequestration%20of%20evergreen%20forest%20in%20Vietnam.pdf. Accessed 27 July 2016

Benjamin K, Sovacool BK (2008) Valuing the greenhouse gas emissions from nuclear power: a critical survey. Energy Policy 36(2008):2940–2953, Elsevier

BP Statistical Review of World Energy (2007) http://www.bp.com/content/dam/bp-country/en_ru/documents/publications_PDF_eng/Statistical_review_2007.pdf. Accessed 26 July 2016

BP Statistical Review of World Energy (2015) https://www.bp.com/content/dam/bp/pdf/energy-economics/statistical-review-2015/bp-statistical-review-of-world-energy-2015-full-report.pdf. Accessed 26 July 2016

CIA (2016) The World Factbook East & Southeast Asia: Indonesia. https://www.cia.gov/library/publications/the-world-factbook/.../countrypdf_id.pdf. Accessed 28 July 2016

Countries of the World (2009) GDP – per capita (PPP) 2007 Country Ranks, 15 Sept 2009. http://www.photius.com/rankings/economy/gdp_per_capita_2007_0.html. Accessed 30 July 2016

Global Footprint Network (2010) Ecological Footprint Atlas 2010, 13 October 2010. http://www.footprintnetwork.org/images/uploads/Ecological_Footprint_Atlas_2010.pdf. Accessed 28 July 2016

Indonesia Ministry of Energy and Mineral Resources (2012) Analyses on Energy Supply and Demand (KajianSuppy and Demand Energy). http://prokum.esdm.go.id/Publikasi/Hasil%20Kajian/ESDM%20SDE.pdf. Accessed 29 July 2016

Indonesian Statistical Agency (2012) Number of Motorized vehicles by Types, Indonesia 1987–2010 (Source: Indonesia Police Office). file:///G:/TRANSPORTATION/type%20of%20vehicles%20in%20indonesia.htm. Accessed 21 Oct 2012

Karyono TH (2008) Bandung thermal comfort study: assessing the spplicability of an adaptive model in Indonesia. Architect Sci Rev 51(1):59–64

Karyono TH (2013) Indonesia sustainable development: minimising CO2 emissions and EF levels without sacrificing the nation's well-being. In: Proceedings the 14th international conference on architecture and sustainable environment, Syiah Kuala University, Darussalam Banda Aceh, 7, 8, 9 November 2013

Orangutan Foundation International (2016) Indonesian Rainforest Facts, https://orangutan.org/rainforest/rainforest-facts/. Accessed 28 July 2016

Sukarna, D (2012) Energy efficiency and renewable energy in Indonesia, Ministry of Energy and Mineral Resources, Republic of Indonesia, Fukuoka, 12–13 July 2012 viewed at http://energy-indonesia.com/03dge/0120720%20Energy%20Efficiency%20and%20Renewable%20Energy.pdf. Accessed 26 July 2016

The World Bank (2016) Environment, World Development Indicators, Energy production and use. http://wdi.worldbank.org/table/3.6. Accessed 26 July 2016

Wikipedia (2013) Energy in Indonesia. http://en.wikipedia.org/wiki/Energy_in_Indonesia. Accessed 5 Oct 2013

Wikipedia (2016a) Geography of Indonesia, https://en.wikipedia.org/wiki/Geography_of_Indonesia,Accessed 28 July 2016

Wikipedia (2016b) Human Development Index. https://en.wikipedia.org/wiki/List_of_countries_by_Human_Development_Index_(2009). Accessed 28 July 2016

Wikipedia (2016c) Java, https://en.wikipedia.org/wiki/Java. Accessed 28 July 2016

Wikipedia (2016d) List of countries by carbon dioxide emissions per capita. http://en.wikipedia.org/wiki/List_of_countries_by_carbon_dioxide_emissions_per_capita. Accessed 28 July 2016

Wikipedia (2016e) List of countries by GDP (PPP) per capita. http://en.wikipedia.org/wiki/List_of_countries_by_GDP_%28PPP%29_per_capita. Accessed 11 Aug 2013

Wikipedia (2016f) List of volcanoes in Indonesia, https://en.wikipedia.org/wiki/List_of_volcanoes_in_Indonesia. Accessed 28 July 2016

Chapter 3
Climate Change and the Built Environment in the Tropics – Is Carbon Enough to Assess Human Impact?

Robert Vale, Brenda Vale, and Tran Thuc Han

Abstract The counting of carbon emissions has become relatively common but it does not reveal the full extent of the environmental impact of human activities, including the effect of the built environment. This article uses the Ecological Footprint concept to explore the effect of the growth-based current economic system on the natural environment. Economic and environmental comparisons are made between a number of "developed" and "developing" countries, revealing that many countries in the tropics have relatively low environmental impact compared to the "developed" world.. However current policies are tending to lead to more environmental damage. The paper uses examples from Vietnam to demonstrate the ways in which many current trends and policies in energy, housing, transport and food are leading to increased environmental impact. It concludes that it may not be possible to reduce impacts without changing the world's current growth-based economic system.

Keywords Ecological footprint • Limits to growth

1 Carbon and Ecological Footprint

The role of human-related carbon emissions in climate change has been well accepted for some time (NRC 2010). Some form of carbon counting is widely used for determining climate change impact, with calls for "carbon-neutral" or "zero-carbon" design; see for example, the United Kingdom's Zero Carbon Hub, established in 2008 as a not for profit organisation to help deliver the UK government

R. Vale (✉) • B. Vale
School of Architecture and Design, Victoria University of Wellington, Wellington, New Zealand
e-mail: robert.vale@vuw.ac.nz

T.T. Han
Energy and Environmental Engineering Research Unit, Faculty of Technology, University of Oulu, Oulu, Finland

T.H. Karyono et al. (eds.), *Sustainable Building and Built Environments to Mitigate Climate Change in the Tropics*, DOI 10.1007/978-3-319-49601-6_3

target of zero-carbon homes by 2016 (Zero Carbon Hub 2014a, b). The United Nations' Intergovernmental Panel on Climate Change (IPCC) recently estimated that in order to have some chance of keeping total average warming below a level at which there could be severe consequences, the world's human population can emit no more than a total of one trillion tonnes of carbon dioxide. Half of this has already been emitted. To meet this limit on total emissions would require reducing greenhouse-gas emissions by 40–70% by 2050 and more or less eliminating them by the end of the century (Talbot 2014). This gives an idea of the scale of the problem.

Carbon alone, in spite of its effects on the global atmosphere, does not reveal the full impact of human life on the resources and systems of the planet Earth. To take account of the full impact of human activities, Wackernagel and Rees in Canada developed the concept of the Ecological Footprint (EF) (Wackernagel and Rees 1996). The EF is a measure of the impact of human activity on the natural environment. More precisely it is the area of bioproductive (ie biologically productive, not desert or built-on) land and water that is required to provide goods and services in a sustainable manner (Ewing et al. 2010, p8). In 2007 the world's average Ecological Footprint per person was 2.7 global hectares (gha – a hectare of land of average productivity) but the available bioproductive land and sea area, the "fair share EF", was only 1.8gha per person (Ewing et al. 2010, p18). This difference between the available area and the area actually used, known as the "overshoot", is made possible by using finite resources such as oil, coal, natural gas, uranium, limestone etc. that are mined from the earth's crust. While it is possible to discover new reserves of oil, gas and other finite resources they are, by definition, finite and will not be available for ever. When these are exhausted all resources will have to be obtained either from the equally finite area of land and water or from the recycling of resources already extracted, there will be no other option. According to data in Ewing et al. (2010, Fig 3, p18) the carbon related part of the global Ecological Footprint in 2007 was only 53% of the total footprint, making clear that to count only carbon is missing much of the problem of human effects on the environment.

What is valuable about the EF compared with other methods of assessing environmental impact, such as carbon counting, is that it makes clear that there are "limits to growth" (see Meadows et al. 1972) because it is based on the Earth's available area of land and water. Once finite resources are consumed the land and sea will have to provide everything, and the area of these is fixed. Land area cannot be increased. The same situation will arise if consumption of finite resources such as carbon-based fuels has to be halted in order to avoid severe climate change, as stated recently by the Executive Secretary of the United Nations Framework Convention on Climate Change (Figueras 2013; Vidal and Readfearn 2013). The situation becomes worse with each passing year as the global population increases and the per capita demand for resources rises with increasing consumer demand.

Ecological Footprint assessment reveals that there is a "fair share" footprint, which is the available area of biologically productive land and sea, the biocapacity, divided between the whole population of the Earth (Vale and Vale 2013). Each year this available "fair share" gets smaller. To illustrate this, in 1961 the fair share

Table 3.1 The change over time in global fair share Ecological Footprint

Date	Population[a]	Fair share EF[b]
1961	3084 million	3.2gha
1971	4089 million	2.7gha (this was the year that consumption exceeded biocapacity)
2007	6631 million	1.8gha
2014	7176 million	?

[a]Population data from US Census Bureau (2013)
[b]Global Footprint data from Global Footprint Network (2012)

footprint was about 3.2gha (Global Footprint Network 2012). In 2007 it was 1.8gha and now is likely to be lower than this. Table 3.1 shows how the fair share has changed since 1961.

To put these fair share footprint figures into the context of a possible standard of living, in 1961 the entire population of the world could have enjoyed the current lifestyle of Mexico, Hungary or Brazil, whereas in 2007, the latest year for which detailed footprint calculations are available, the whole world could live like the current populations of Myanmar, Ghana or Armenia (based on EF data from Ewing et al. 2010, Fig 7, pp 19–20).

2 Economics and Growth

Viewed from an economic perspective the world appears much better off on average. The world Gross Domestic Product (GDP) per capita has gone from US$5323 per capita in 1995 (the earliest year for which the World Bank publishes per capita GDP data) to US$10,514 in 2013 (World Bank 2014a). This current average GDP means that the entire population of the world could have the current living standards of Malaysia, Mexico or Turkey. Herein lies the problem. In the realm of economics, the type of life to which people can aspire has no limits. It is assumed that income can continue to grow, although it is worth noting that growth is not evenly distributed. It is also not entirely clear if progress is being made. According to the World Bank's GDP data, in 1995 the entire world could have lived like people in the Czech Republic at that time (data from World Bank 2014a). To show the effect of economic growth, Table 3.2 gives some figures for the growth in per capita income in the developed economy of the United States and the developing economies of India and China. The income figures quoted come from the US Census and the OECD and are all given in 1996 dollars to allow comparison (Vale and Vale 2009, pp 21–22).

Table 3.2 shows that in the thirty years from 1972 personal incomes in India and China have grown much faster than those in the United States, with China's rate of increase being roughly ten times the increase in United States income over the same period. This would appear to support the view that the global economy is bringing people out of poverty. The figures in Table 3.2 can also be used to show a different situation. Table 3.3 shows the gap in average per capita income between each of the two "developing" countries and the United States, again expressed in 1996 dollar values.

Table 3.2 Per capita income growth in some rich and poor countries

Country		US$ 1996 value	
United States of America			
1972	US disposable personal income	$13,692	
2002	US disposable personal income	$24,479	
Increase over thirty years		79%	
India			
1972	India per capita income	$ 1100	8% of US income in 1972
2002	India per capita income	$ 2700	11% of US income in 2002
Increase over thirty years		145%	
China			
1972	China per capita income	$500	4% of US income in 1972
2002	China per capita income	$4000	16% of US income in 2002
Increase over thirty years		700%	

USA data from http://www.census.gov/statab/hist/HS-33.pdf (viewed 18 March 2008)
Per capita income trends in China and India 1972–2002 based on data from the Organisation for Economic Cooperation and Development given in Pucher J. et al (undated) *Urban Transport in India and China: Slide No. 3* University of Sydney and Rutgers University. Available at http://www.itls.usyd.edu.au/seminars/presentations/johnp.pdf (viewed 18 March 2008) and at http://sydney.edu.au/business/__data/assets/pdf_file/0003/28749/johnp.pdf (viewed 1 December, 2014)

Table 3.3 Per capita income gap for India and China compared with the United States

Country		US$ 1996 value
India		
1972	Income gap with United States	$12,592
2002	Income gap with United States	$21,779
China		
1972	Income gap with United States	$13,192
2002	Income gap with United States	$20,479

On the basis of the data in Tables 3.2 and 3.3, personal incomes in India and China still fall far behind those of the developed world. Incomes in 2002 in these two countries are not even close to what incomes were in the USA in 1972, and the income gap between both of the poorer countries and the wealthy one has nearly doubled in both cases over the 30 years between the two sets of data. Economic growth is indeed making the poor richer in absolute terms as people in both India and China are on average better off than they were 30 years ago. At the same time economic growth seems to be having the effect of making people in these two countries significantly poorer in relative terms. The poor indeed have got richer, but at the same time the rich have got much richer meaning that the poor have become relatively poorer. The World Bank, the source of the data used in the discussion above, states that its "mission" is to "... boost shared prosperity" (World Bank 2014b). This type of economic world-view drives the world economy in its current form and for many people it is an attractive way to view the world (Local Government New Zealand 2014). This world-view says that we can all have more prosperity and that there are no limits to growth. The World Bank speaks very positively of economic growth, stating in its *Global Economic Prospects* report that "Despite the early weakness, growth is expected to pick up speed as the year progresses and world GDP is projected to expand by 3.4 percent in 2015 and 3.5 percent in 2016 –broadly in line with earlier forecasts" (World Bank 2014c p3). Similarly, the government of the United Kingdom states "To make sure the UK can succeed in the global economy, we are taking action to stimulate economic growth..." (HM Treasury 2013).

3 Limits to Growth

Although economists regard continuous growth as essential, some scientists argue that natural laws run counter to the idea of continuous growth. For example the Associate Director at the Center for Astrophysics and Space Sciences of the University of California asks "Can economic growth last?" and goes on to explain why the laws of physics make continued economic growth impossible (Murphy 2011). It is worth pointing out that the laws of physics are fixed by the physical world whereas the "laws" of economics are entirely a human construct, they are made up to fit a set of beliefs or theories. Unlike in economics, the natural world does not provide examples of things growing continuously, for example a mouse never grows to become an elephant. There is growth in nature but it stops when the appropriate size is reached. Phenomena that do grow continuously, such as cancer, generally end up being fatal for their host. In 1972 a group of scientists commissioned by the Club of Rome, a global "think-tank" founded in 1968 to think about the future of humanity (Club of Rome undated), published a report called "The Limits to Growth" (Meadows et al. 1972). This used computer modelling of population, food, industrialisation, resources and pollution to predict what would happen to human civilisation in the future. The conclusion was that if current policies of unrestrained growth were continued, civilisation would collapse somewhere around the middle of the twenty-first century. The Limits to Growth was widely criticised, particularly by economists (see Cole and Pavitt 1973 and the study by Bardi 2011) and has been described as being consigned to the "dustbin of history" (Lomborg and Rubin 2002).

More recently a scientist at the government Commonwealth Scientific and Industrial Research Organisation (CSIRO) in Australia carried out a study which compared the predictions made in the Limits to Growth modelling with what had happened in reality between 1970 and 2000 (Turner 2008). He found that reality had very closely followed the path predicted by the Limits to Growth model and concluded "...the observed historical data for 1970-2000 most closely matches the simulated results of the LtG (Limits to Growth) "standard run" scenario for almost all the outputs reported; this scenario results in global collapse before the middle of this century"(Turner 2008, p37). Turner's more recent work on the same study suggested that the collapse would start in 2015, with global industrial output falling

Table 3.4 2007 Ecological Footprints of a selection of "developed" and "developing" countries (rounded)

The "West"	
United States of America	8.0gha (Ewing et al. 2010, p31)
Australia	6.8gha (Ewing et al. 2010, p31)
United Kingdom	4.9gha (Ewing et al. 2010, p31)
New Zealand	4.9gha (Ewing et al. 2010, p31)
Asia	
Thailand	2.4gha (Ewing et al. 2010, p29)
China	2.2gha (Ewing et al. 2010, p29)
Fair share EF	**1.8gha**
Vietnam	1.4gha (Ewing et al. 2010, p28)
Indonesia	1.2gha (Ewing et al. 2010, p29)

suddenly, followed by falls in food supplies and services (Turner 2012). Out of self-interest if not altruism it would seem advisable for nations to follow policies that would be likely to avoid this mid-century collapse.

Table 3.4 shows the range of Ecological Footprints for some "developed" and "developing" countries. The values are for 2007 as this is the last date for which data are available at the time of writing.

It is clear from Table 3.4 that the countries in Asia have much lower footprints than any of the "developed" countries, and two of the tropical countries in this table, Vietnam and Indonesia, have footprints that are well below the fair share. The fair share is important in this discussion because of the limited availability of bioproductive land and sea. In the economic world-view everyone can have more and there are no limits to growth. If, on the other hand, science is correct and continuous growth is not possible, the current economic world-view will have to be abandoned. In a world-view based on Ecological Footprint the only way that one nation can develop and achieve its fair share of the world's resources is for a country which has more than its fair share to have less. In this scenario Vietnam and Indonesia have some advantages. It is probably easier to develop up to a point and then level off than it is to reverse development as will have to happen in the "developed" countries. However there is little sign that tropical countries are tending to follow policies that will lead to stabilized Ecological Footprints. On the contrary, they seem to be following the example of what might be called from an EF perspective the "over-developed" countries.

4 Policies and Footprints

A recent study of the city of Hanoi, the largest city in Vietnam, has shown that the current low Ecological Footprint is the result of a number of patterns of behaviour that are different from those of the developed world (Tran 2014). These include the fact that people live in larger households sharing smaller houses, they have smaller vehicles (typically motorcycles) and also walk and cycle more, they also eat less meat, dairy and non-basic foods and do not waste food (Tran, op cit). However currently proposed policies and trends are likely to result in an increase in footprint. The following sections, summarised in Table 3.5, will look at these policies and trends and their likely effects in the areas of energy, housing, personal transport and food.

4.1 Energy

Published energy targets (International Resources Group 2011), show that the total amount of renewable energy used in Vietnam is planned to be less in 2030 than in 2008, although the use of individual types of renewable energy (hydro, wind and solar) will increase. The result is that the energy component of the Vietnam Ecological Footprint increases by 62% by 2030. At the same time the use of non-renewable energy (and the resulting carbon emissions) is planned to increase by more than 200% by 2030. This is not a sensible strategy in a world which needs to reduce emissions.

4.2 Housing

According to a recent study (Nguyen et al. 2010), by 2030 household size in Vietnam is projected to decline from 4 to 3.5 people/house. The total number of households in Vietnam in 2005 was about 20 million and is projected to increase to 30 million by 2030. The total per capita housing Ecological Footprint of Hanoi is projected to increase by up to 35% due to the decline in household size and depending on the cooling method (fan or air conditioning). This increase in footprint might be not significant at the individual level, but it would be a significant increase in the total housing footprint of the whole country with its population of 100 million people by 2030.

4.3 Personal Transport

Hanoi suffers from lack of a public transport system, with serious congestion problems, a rapid increase in car ownership and mass development of new residential areas without provision for local jobs, public services, schools, health centres, etc. (World Bank 2006; Nguyen 2007; Labbé 2010). As the city grows, travel distances to work, public services, and education become less suitable for walking and cycling, and people have to rely more on private transport due to the poor public transport system. Furthermore, economic growth means more people would be able

Table 3.5 The effect of current and proposed patterns of living, housing, transport and food on individual Ecological Footprint, Hanoi, Vietnam

Issues	Current patterns of living	Current Ecological Footprint (gha/person/year)	Proposed patterns of living	Future Ecological Footprint (gha/person/year)
Energy			**Year 2030**	
	Non-renewable	**0.26**	**Non-renewable**	**0.42**
	1089PJ		**3418PJ**	
	Renewable		**Renewable**	
	1089PJ		**936PJ**	
Housing			**Year 2030**	
Low-user	Small house	0.052	**Low-user:** less sharing	0.055
	More sharing			
	4 persons per house		3.5 persons per house	
	19m² per person		Over 21m² per person	
	Electric fan cooling		Electric fan cooling	
High-user	More sharing	0.066	**High-user:** less sharing	0.070
	4 persons per house		3.5 persons per house	
	19m² per person		Over 21m² per person	
	Air conditioner cooling		Air conditioner cooling	
Transport			**Year 2020**	
Fossil fuels	11,008km/person/year	0.19	**"Business as Usual"**	0.32
			16,466km/person/year	
	Small private vehicle, more walking and cycling		Increase car travel less cycling and walking	
			Remain share of public transport	
	Non-motorised: 30%		Non-motorised: 9%	
	Bus: 2%		Bus: 2%	
	Motorcycle: 62%		Scooter: 53%	
	Car: 2%		Car: 36%	
			"High Mass Transit"	0.27
			13,433km/cap/year	
			Increase car travel	
			Less cycling and walking	
			More public transport	
			Non-motorised: 14%	
			Bus and train: 4%	
			Motorcycle: 47%	
			Car: 35%	

(continued)

Table 3.5 (continued)

Issues	Current patterns of living	Current Ecological Footprint (gha/person/year)	Proposed patterns of living	Future Ecological Footprint (gha/person/year)
Food			**Year 2050**	
	414kg/cap/year	0.55	590kg/cap/year	0.85
	Less meat and dairy products, very little non-basic food consumption		Double amount of meat and dairy consumption, sweets and alcohol as the current level of Wellington, NZ	
	Meat and dairy: 55kg/cap/year		Meat and dairy: 104kg/cap/year	
	Non-basic food: 31kg/cap/year		Non-basic food: 158kg/cap/year	
TOTAL	**Low-user house, current energy and transport fuels, and current food consumption**	**1.05**	**Scenario having smallest footprint - Low-user, High Mass Transit**	**1.59 (Ecological Footprint increase 51%)**
TOTAL	**High-user house, current energy and transport fuels, and current food consumption**	**1.07**	**Scenario having largest footprint - High-user, Business as Usual**	**1.66 (Ecological Footprint increase 55%)**

to afford to own motorcycles and cars, meaning fewer people would cycle and walk. Schipper et al. (2008 p.14), give two scenarios for 2020. The first is Business as Usual (BAU), which adopts current trends in transport. The share of public transport by bus will continue to increase from 6% of all trips to work per day to 14.5% in 2020. The second scenario, High Mass Transit (HMT), recommends installing urban rail rapid transit and upgrading the urban bus capacity. Under this second scenario, the public transport share of total trips would reach 30%.

As recently as 1995 Hanoi was still a cycling and walking city with fewer private vehicles. By 2005 the increasing dominance of private vehicles was seen, although walking and cycling still accounted for nearly a third of the total distance travelled. In 2020 even the High Mass Transit scenario assumes more distance travelled overall, a large increase in private car use and a large decrease in cycling. The results show that transport in Hanoi would have much higher footprints in both the scenarios proposed for 2020, more than double for business as usual and still 1.7 times higher even if the public transport system is upgraded with urban rapid rail. The key reason is the significant proposed increase in private transport use, in particular car use, and the decline in non-motorised transport by 2020. In order to move towards a more sustainable system of transportation, policy would need to limit the use of private transport, especially car use, for example eliminating private vehicles in the city centre by promoting walking, cycling and public transport, as well as providing a better mass transit network as proposed in the HMT scenario.

4.4 Food

Food is another key area for reducing carbon emissions (Bajželj et al. 2014). Increasing consumption of meat and dairy products and other high EF foodstuffs is leading to increased emissions, to the point where it is predicted that the global food system alone could reach the global target for greenhouse gas emissions (University of Cambridge 2014). Although the current food consumption in Vietnam is much lower than that provided by Western diets, according to FAO predictions (Alexandratos 2006 p.26), by 2050 the total calorie intake per person per day in East Asia (including Vietnam) would increase from 2800kcal/day to over 3200kcal/day. The FAO report predicted that this increase would largely come from the consumption of meat and dairy products which is expected to double by 2050. In addition, it is likely that consumption of non-basic foods such as sugar and alcohol will also increase to western levels and this will also drive up the Ecological Footprint.

5 Conclusions

It is clear from the figures in Table 3.5 that government policies and likely consumer and other trends will result in an increased Ecological Footprint for Hanoi. This also means increased carbon emissions. This does not seem to be a good outcome in the long term. Similar current policies and consumer trends based on economic growth are found in all countries, both those in the tropics and those which are already highly developed, arguably over-developed in terms of their Ecological Footprint. This means that there is no chance that demand for finite resources will cease to increase. Continued economic growth and population growth make it highly likely that carbon emissions will also continue to increase. The conclusion is that it may not be possible to reduce emissions until the world changes its economic focus.

References

Alexandratos N (2006) *World agriculture: towards 2030/2050* Interim Report. Rome: Global Perspective Studies Unit, Food and Agriculture Organization of the United Nations.

Bajželj B, Richards KS, Allwood JM, Smith P, Dennis JS, Curmi E, Gilligan CA (2014) Importance of food-demand management for climate mitigation. Nat Clim Chang 4:924–929

Bardi U (2011) *The limits to growth revisited*. Springer, New York

Club of Rome (undated) About the Club of Rome. Available at http://www.clubofrome.org/?p=324. Viewed 21st Nov 2014

Cole H, Pavitt K (eds) (1973) *Models of doom: a critique of the limits to growth*. University of Sussex, Brighton

Ewing B, Moore D, Goldfinger S, Oursler A, Reed A, Wackernagel M (2010) *The ecological footprint atlas 2010*. Global Footprint Network, Oakland

Figueras C (2013) Keynote address. Warsaw, Poland: World Coal Association International Coal & Climate Summit, 18 November, 2013

Global Footprint Network (2012) "World" 5th August. Available at http://footprintnetwork.org/en/index.php/GFN/page/trends/world/. Viewed 20th Nov 2014
HM Treasury (2013) Policy: achieving strong and sustainable economic growth. 31 July, Available at https://www.gov.uk/government/policies/achieving-strong-and-sustainable-economic--growth Viewed 3 December 2014
International Resources Group (2011). Energy trends in developing Asia: priorities for a low-carbon future (Annex I: Country data and figures). United States Agency for International Development
Labbé D (2010) Facing the urban transition in Hanoi: recent urban planning issues and initiatives. Institut National de la Recherche Scientifique, Centre - Urbanisation Culture Société, Quebec
Local Government New Zealand (2014) LGNZ advocates for the need for economic growth across all of New Zealand. 11th August 2014 available at http://www.lgnz.co.nz/home/news-and-media/2014-media-releases/lgnz-advocates-for-the-need-for-economic-growth-across-all-of--new-zealand-/ Viewed 21st November, 2014
Lomborg B, Rubin O (2002) The dustbin of history: limits to growth. Foreign Policy 133:42–44
Meadows DH, Meadows DL, Randers J, Behrens W (1972) *The limits to growth*. Universe Books, New York
Murphy T (2011) Can economic growth last? In: Do the math. 14th July available at http://physics.ucsd.edu/do-the-math/2011/07/can-economic-growth-last/ Viewed 21st November, 2014
Nguyen HT, Gomi K et al (2010) A low carbon society development towards 2030 in Vietnam. Kyoto University, Asia Pacific Integrated Model, National Institute of Environmental Studies, Kyoto
Nguyen QN (2007) Integration of transport and land use in Hanoi: can we relieve traffic congestion by relocating some major land-uses? Master of Science in Urban Planning and Land Administration. International Institute for Geo-Information Science and Earth Observation, Enschede
NRC (2010) Advancing the science of climate change. National Research Council. Washington, DC: The National Academies Press
Schipper L, Anh TL et al. (2008) Measuring the invisible: quantifying emissions reductions from transport solutions. Available at http://pdf.wri.org/measuringtheinvisible_hanoi-508c_eng.pdf. Viewed 12th December, 2011
Talbot D (2014) Climate panel issues dire report as renewables make little impact. MIT Technology Review. November 3 Available at http://www.technologyreview.com/news/532251/climate-panel-issues-dire-report-as-renewables-make-little-impact/. Viewed 10th December 2014
Tran TH (2014) Sustainable patterns of living based on an investigation of footprint in Hanoi-Vietnam, Wellington-New Zealand and Oulu-Finland. PhD thesis, Wellington, New Zealand, Victoria University of Wellington
Turner G (2008) A comparison of the limits to growth with 30 years of reality, Socio-economics and the environment in discussion CSIRO working paper series 2008–09, CSIRO, Canberra
Turner G (2012) On the Cusp of global cllapse? Updated comparison of the limits to growth with historical data. GAIA 21/2:116–124
University of Cambridge (2014) Changing global diets is vital to reducing climate change. Research 1st September. Available at http://www.cam.ac.uk/research/news/changing-global-diets-is-vital-to-reducing-climate-change. Viewed 12th December, 2014
US Census Bureau (2013) International programs: international data base. Available at http://www.census.gov/population/international/data/idb/worldpoptotal.php. Viewed 20th November 2014
Vale R, Vale B (2009) Time to eat the dog? The real guide to sustainable living. Thames and Hudson, London
Vale B, Vale R (eds) (2013) Living within a fair share ecological footprint. Routledge, London
Vidal J, Readfearn G (2013) Leave coal in the ground to avoid climate catastrophe, UN tells industry. London, England: The Guardian, Monday 18 November 2013. Available at http://www.theguardian.com/environment/2013/nov/18/leave-coal-avoid-climate-catastrophe-un Viewed 13 Nov 2014

Wackernagel M, Rees W (1996) Our ecological footprint. New Society Publishers, Gabriola Island
World Bank (2006) Project brief on a proposed credit in the amount of USD 176 million equivalent and a proposed grant from the Global Environment Facility Trust find in the amount of USD 9.8 million equivalent to the Socialist Republic of Vietnam for a Hanoi Urban Transport Development Project. The World Bank, Washington, DC
World Bank (2014a) GDP per capita (current US$). Available at http://search.worldbank.org/all?q term=global+gdp+per+capita+history&title=&filetype=. Viewed 20th November 2014
World Bank (2014b) Mission. Available at http://www.worldbank.org/en/about viewed 28th November 2014
World Bank (2014c) Global economic prospects, volume 9, June 2014. Washington, DC: World Bank Group
Zero Carbon Hub (2014a) What we do. Available at http://www.zerocarbonhub.org/what-we-do, viewed 13 Nov, 2014
Zero Carbon Hub (2014b) Zero carbon homes and nearly zero energy buildings: UK Building Regulations and EU Directives. Zero Carbon Hub, London

Chapter 4
Monitoring Climate Change Adaptation: Lessons from Scotland

Susan Roaf and Katherine Beckmann

Abstract The growing scale and frequency of the consequences of climate change are writ large across local, regional and international landscapes and increasingly require governments to develop policies and action plans that can effectively reduce and manage the impacts of more extreme weather on their social, economic and physical systems. In order to Mitigate, one has to measure how well policies to effect changes in such national systems are working, how well, in fact, they are resulting in successful adaption to an altered world. Scotland has pioneered the development of its own locally appropriate Adaptation Indicators within the framework of a Scottish Climate Change Adaptation Programme (SCCAP). This has been funded to address the impacts identified for Scotland in the first UK Climate Change Risk Assessment (CCRA). A first set of indicators developed relates to the impacts on populations of extreme weather and climate events in the contexts of the built and natural environments. As the hazards posed by a changing climate manifest themselves and escalate, regions are preparing action plans to protect themselves against those growing impacts over time. There is an important role to be played here by built environment professionals in the development of such national indicator sets and the implementation of the related climate mitigation and adaptation action plans. This paper sets out how we in Scotland approached the challenge of developing related indicator sets, some of the lessons learnt in framing them and challenges identified in doing so.

Keywords Adaptation • Indicators • Climate change

S. Roaf (✉) • K. Beckmann
EGIS, Heriot Watt University, Edinburgh, UK
e-mail: s.roaf@hw.ac.uk

T.H. Karyono et al. (eds.), *Sustainable Building and Built Environments to Mitigate Climate Change in the Tropics*, DOI 10.1007/978-3-319-49601-6_4

1 Introduction

Rising to the challenge of 'future-proofing' our built environment against climate change, Scotland and the UK, like many other countries, are undertaking a programme of climate adaptation planning and 'Informed Decision Making'. In order to '*Mitigate*', one has to measure how well policies to effect changes in the physical, environmental, economic and social conditions of a nation are working, how well, in fact, they are adapting to an altered world. To do this requires a focus on deliberative policy making to formulate both Policy Targets and Action Plans aimed at:

(a) Reducing our exposure and vulnerability to the risks inherent in a changing climate.
(b) Exploring ways of exploiting any opportunities that may emerge with climate change.
(c) Developing the sound adaptation policy framework that is increasingly understood as being critical to the encouragement, enablement and implementation of planned action plans designed to enhance the resilience of populations over time.

This requires an evolutionary process to be established that can accommodate timely adjustments, as and when necessary, to enable populations to cope with, and adapt to, new landscapes, buildings and lifestyles in an increasingly rapidly changing climate.

However, our understanding of adaptation is still rudimentary. We are only just beginning to implement related policies and understand how such adaptive processes might work in practice. First of all we need to know whether the actions we take now are contributing to desired outcomes: Are we doing the right things? Are they in the right places? Is there enough action? Is policy encouraging the right kinds and levels of adaptation?

Functionalising feedback loops between the policy, the regulatory and the multiple systems at play is key to the success of an adaptive policy process. It is essential to use compatible monitoring metrics and measures within the system to link evolving political policies to their impacts 'on the ground'. Much work has been done to date on this subject spring-boarded from the initial UK Climate Change Risk Assessment Report (DEFRA 2012). The Scottish Government has developed an integrated climate resilience programme (Scottish Government 2014) that includes an over-arching set of Adaptation Indicators. The Scottish Government worked with its agencies, academics and other experts to develop indicators to monitor climate change adaptation in Scotland, covering the natural environment (agriculture, biodiversity and forestry), buildings and infrastructure and society themes represented in the Scottish Climate Change Adaptation Programme (SCCAP). To monitor adaptation, indicators are used that track change in adaptation actions and in climate change risks and opportunities and their associated impacts.

The Adaptation Indicators Framework measures and monitors the direction of travel of a related policy towards improving or worsening the case at hand, and the magnitude of its success or failure can be measured and in turn inform new policy directions and the strengths of the policy levers applied within them. The traditionally long timescales involved in many adaptation impacts may appear to be less useable in informing current policy because they provide an evidence base for change over a far longer term than that over which present day decision makers can have influence. We now live in what is increasingly understood to be a non-linear climate system in which even decadal impacts may, or may not, provide vital information on the nature and rates of change within the systems involved.

Frequently asked questions are what is *Adaptation*? What is *Successful Adaptation*?

2 Concepts Underlying the Notion of Adaptation

Models of climate change impact, adaptation, and of indicator development are considered in a number of sources. They rely on the inter-linked concepts of hazard, exposure, sensitivity, risk, vulnerability, resilience, impacts, adaptive capacity and adaptive action. The reports of the Intergovernmental Panel on Climate Change (IPCC) provide an important reference point to these. Working Group II of the IPCC (2007a, b) worked to provide a framework to 'assesses the vulnerability of socio-economic and natural systems to climate change, negative and positive consequences of climate change, and options for adapting to it'. In the IPCC's Third Assessment Report (TAR) this group focused on impacts, adaptation and vulnerability (IPPC 2001). The report explored some of the key concepts in the field of adaptation, such as adaptive capacity, vulnerability and resilience, as well as providing definitions of them. The ways in which these terms are used, and their definitions, vary between authors and reports, to the extent that both guidance reports and journal papers have been produced on the topic (Hofmann et al. 2011). A report on key terms in adaptation produced by the OECD provides a useful reference guide (Levina and Tirpak 2006).

Adaptive Capacity was an idea developed in the IPCC's 2001 report. Adaptive Capacity at a national level is determined by factors such as economic wealth, technology and infrastructure, information and knowledge, and equity and social capital. Adaptive capacity is an important concept because trends, drivers and policies that affect it will enable (or hinder) the ability to carry out planned adaptation. In the literature explored, both the determinants of and indicators for Adaptive Capacity were given at a country level, rather than at a sectoral or individual risk level (Metzger et al. 2006). It has been noted by various authors that Adaptive Capacity per se may not necessarily translate into adaptation (Berrang-Ford et al. 2011). Nevertheless, the concept is worth retaining, as actions that increase Adaptive Capacity may enable a greater range of future options than actions that are focused on more rigid adaptation plans. Thus it follows that indicators that measure Adaptive Capacity (or the related concept of resilience) may be more appropriate in some instances than indicators that monitor fixed plans (Engle 2011).

Vulnerability was also defined within the IPCC's Third Assessment Report which states that the degree of vulnerability to changes in climate (hazard) is determined by a subject's exposure, sensitivity and ability to adapt to that change (Eriksen and Kelly 2007). A later IPCC definition is given that refines this definition and states that 'Vulnerability is a function of the character, magnitude, and rate of climate change and variation to which a system is exposed, and the sensitivity and adaptive capacity of that system' (IPPC 2007a and b).

Vulnerability will differ for each system under consideration (not just each sector, but each impact within a sector will have different contributors to vulnerability). Developing indicators of vulnerability will therefore require a large number of indicators, which may not be immediately suitable for policy or communication needs. Nonetheless, the concepts of function or system maps, for both adaptive capacity and for vulnerability, are important as they may show policy makers and practitioners *where best to focus policy and action*. It is likely that professionals in any given field will have implicit knowledge of such systems, and an intuitive sense of where action is best focused. This is one of the many reasons why indicators are best devised in close collaboration with stakeholders and professionals.

Resilience, often seen as the converse of vulnerability, was initially defined by the IPCC as the amount of change a system can undergo without changing state. This definition has changed, and resilience is now defined as 'the ability of a social or ecological system to absorb disturbances while retaining the same basic structure and ways of functioning, the capacity for self-organisation, and the capacity to adapt to stress and change'. This is similar to the definition given by the UK Climate Impacts Programme (UKCIP) which is 'the ability to recover from (damaging) effects'. Optimising the resilience of a system is an important concept in adaptation (and for indicator development), as it does not aim to be prescriptive about the exact mechanisms of adaptation.

Risk is another important concept for adaptation, and the UK Climate Change Risk Assessment defines risk as 'combining the likelihood an event will occur with the magnitude of its consequences' (DEFRA 2012). Interplay between the *drivers of risk* is also depicted in a 'risk triangle' (Crichton 2001), where the interaction of the three external factors of hazard, vulnerability and exposure determines the actual risk (or impact) experienced.

In considering frameworks for adaptation and indicator development it is important to have a common understanding of the terms used, and an agreed glossary may be one way to achieve this.

Adaptation Strategies range through preventing or accepting losses, to spreading costs through to shifting activity or location to avoid problems. There is not necessarily one 'best' fit for any given situation – the approach chosen will depend on the priorities and values of those involved.

3 Different Classes of Indicators

Different indicator sets have been proposed and developed for several of the concepts introduced above although the underlying phenomenon is of course climate change and *Climate Change indicators* measure *actual* changes in climate and environment. Climate change creates *impacts* that affect the environment, society and the economy. *Climate Impact indicators* measure these impacts. For example, temperature-related impacts include electrical outages and cooling equipment purchases. Rainfall related impacts include sewer overflows and flooding.

Vulnerability indicators measure the degree to which a system is susceptible to, and unable to cope with, the adverse effects of climate change.

*Resilience indicators*measure the amount of change a system can undergo without changing the functioning of the underlying systems. A framework using indicators (of the characteristics) of a resilient natural environment is being proposed by Natural England in their work towards indicators of adaptation (Natural England 2010). Resilience Indicators are indicators used for the natural environment and for some other systems. As the name suggests, these give an indication of the resilience of a system. An examples is 'Bird Population Numbers' (as a proxy for ecosystem resilience) as proposed by Natural England and used within the UK Sustainable Development Indicators (Natural England 2010).

*Adaptive Capacity indicators*measure the capacity or potential for adaptation (rather than the actual adaptation). and are well developed in a Tyndall centre publication in 2004.

Adaptation Indicators measure our *adaptation* – ie. how well we are equipping ourselves to deal with such impacts, and how resilient our systems are in responding to such climate-driven impacts concerning 'adjustment in natural or human systems in response to actual or expected climatic stimuli or their effects, which moderates harm or exploits beneficial opportunities' (IPPC 2007a and b).

Process indicators show performance in *building adaptive capacity*. Such indicators can monitor the progress in implementing adaptation measures but do not report on actual outcomes. They report on actions that are generally pro-active. An example is the 'Number of Land Management Plans that take Climate Change into account', Proposed by Natural England in 'Adaptation Indicators for the Natural Environment' (Natural England 2010).

Outcome indicators show performance in *delivering adaptation actions and outcomes*. Such indicators measure the results of adaptation policies and the effects of wider activities. Progress, or deterioration in progress, may not be visible for some years. An example is the 'Value of settled insurance claims for weather perils', from the set of 34 climate change indicators published by DEFRA (2004))

Sector specific indicators focus on performance within a particular sector and an example is the 'Installation of flood resistance measures in high-risk homes' as used in the report 'Adapting to Climate Change in the UK: Measuring Progress' (Adaptation Sub-Committee 2011).

Cross-cutting indicators measure outcomes that build on action in more than one sector and/or more than one level. These may also act as *Headline indicators* which are based on outcomes that are contingent on performance across several different sectors. Examples include 'Seasonal Pattern of Human Mortality' and 'Level of settled adverse weather claims per year' from the indicator set first published by DEFRA in 1999 and reviewed in 2003 (DEFRA 2004).

Other characteristics to consider include aspects such as whether indicators are able to account for open-ended adaptation, as opposed to pre- defined (prescriptive) measures. The latter may be easier to measure (e.g. number of new developments incorporating Sustainable Urban Drainage Systems), but they may also drive inappropriate or ineffective adaptation.

4 Climate Change as a 'Wicked Problem'

Climate change is a 'wicked' and potentially devastatingly expensive problem, being chaotically complex in construction (Roaf 2013). It results in potentially catastrophic social, economic and environmental impacts and cuts across traditional policy and decision making contexts, being an eco-system level, cross-sectoral, phenomenon. Adaptation systems contain many different factors which affect change in outcomes. It is also a long term issue and of such a magnitude that it may exceed conventional funding time scales and the financial capacities of local economies. Hurricane Sandy in late October 2012 killed some 54 people and cost the American tax payer over $68 billion dollars, well beyond the budgets of the east coast US states where it wrought its havoc (Wikipedia, Hurricane Sandy). Typhoon Haiyan in early November 2013 that killed far more people, some 100,000 in the Philippines, is reported to have caused far less financial impact, at a cost of some $14 billion in damage (Wikipedia, Typhoon Haiyan). It did however, result in an 8 % drop in the GDP in that country. Large amounts of internationally sourced aid were needed to reinforce the local economy and rebuild devastated areas.

These inherent complexities in evaluating and assessing costs and impacts of such unpredictable extreme events make it difficult to design and implement adaptation policies and related monitoring systems to cope with them. However there are increasingly high resolution models pointing towards, and predicting, the numerous emerging manifestations of the changing climate, which when coupled with historic evidence on climate impacts to date, can be usefully woven together to inform the structure of effective monitoring tools for adaptation policies so reducing the vulnerabilities, if not the exposures, of populations to related impacts. A confounding factor is often the difficulty of untangling the various contributory influences on the scale of climate impacts from the many different drivers of change. Whilst Monitoring and Evaluation (M&E) is commonly used as part of the policy cycle, its application to climate adaptation is new and it provides a new 'science-policy' challenge.

5 The UK Climate Change Risk Assessment Programme (CCRA)

As Climate Change rose up the political agenda the UK initiated a Climate Change Risk Assessment (CCRA) process. The first findings report was presented to Parliament in January 2012. The CCRA has reviewed the evidence for over 700 potential impacts of climate change in a UK context. Detailed analysis was undertaken for over 100 of these impacts across 11 key sectors, on the basis of their likelihood, the scale of their potential consequences and the urgency with which action may be needed to address them.

Scotland is a unique subdivision of the United Kingdom and warranted a tailored CCRA report in which there were the 11 topic-specific reports produced on Agriculture, Forestry, the Built Environment, Energy, Transport, Water, Floods and Coastal Erosion, and reported under the five key themes of:Natural Environment; Agriculture and Forestry; Business and Services; Infrastructure and Buildings; Health and Wellbeing. The CCRA outputs provide an evidence base that can be used by central Government and Devolved Administrations in identifying *priorities for action and appropriate adaptation measures*that will be required to minimise risks to our economy, environment and society.

Some of the stepping stones in the evaluation of the direct and indirect risks that the CCRA identifies for the built environment and infrastructure services include:

- UK Climate Change Act 2008: 5-yearly assessments of risk
- UK First Climate Change Risk Assessment (CCRA) published January 2012
- CCRA of current and predicted threats and opportunities from climate change to 2100

6 The Scottish Climate Change Adaptation Programme (SCCAP)

In Scotland climate change policy is driven by the 2008 Climate Change Scotland Act, which contains a legal requirement for a Climate Change Risk Assessment and a Scottish Climate Change Adaptation Programme. The mission and vision of a *Climate Ready Scotland* is designed to: *To increase the resilience of Scotland's people, environment and economy to the impacts of a changing climate'*.

The first Scottish Climate Change Adaptation Programme was published in May 2014. It identified a range of policies and actions in three sectors: the Natural Environment, Buildings and Infrastructure Networks and Society and was aimed at encouraging and enabling adaptation to the risks highlighted in the CCRA for Scotland. It outlined:

- Objectives in relation to adaptation to climate change
- Proposals and policies for meeting those objectives
- Periods within which those proposals and policies will be introduced

The Scottish Government (SG) invested in a climate Risk, Mitigation and Adaptation research programme in 2010 involving hundreds of researchers across Scotland from all the related sectors. The programme that was managed through the Climate Exchange in Edinburgh (http://www.climatexchange.org.uk) has developed a suite of *adaptation indicators* to facilitate the measurement and understanding of the extent to which, and ways in which, Scotland is adapting to climate change. A political requirement was that the Scottish Climate ChangeAdaptation Programme (SCCAP) should be independently scrutinised by the UK Adaptation Sub-Committee to determine value for money and robustness of the process.

These indicators are derived from a risk based approach as this is the basis of adaptation policy making in Scotland (and the UK) with the SCCAP responding to the Climate Change Risk Assessment. It also fits well with the decision making contexts in which the indicators are to be used as many decision makers are already working with risk management processes.

The Scottish Climate Change Adaptation Programme has achieved to date:

- *56 policies and proposals*
- *Developing Research Programme:* Resilience measures for new and existing buildings and the development of an Adaptation Indicator set for Scotland.
- *Improving data availability:* Central coordination for data on climate effects on transport networks.
- *Knowledge provision:* Sustainability in Architecture programme.
- *Joint working:* Joint Agency programmes to protect historic sites including the Scottish Historic Buildings, Environment, Natural and Cultural organisations.
- *Implementing existing polices:* Secure and Resilient- a Strategic Framework for Critical National Infrastructure.
- *Improving existing policies:* Review of Building Regulations.
- *Proposing new policies:* Eg: New mandatory standards for water efficiency.

Here are some examples of developing *Adaptation Indicators* that have been developed by the Scottish team for the Built Environment including their Risk, Action and Impact dimensions and Sources for the related data):

Indicators for Flooding (Data sources: Scottish Environmental Protection Agency (SEPA); Ordinance Survey Master Maps)

- Risk of flooding to residential and non-residential property and to cultural heritage.
- Risk indicators e.g. proportion of properties/assets at risk
- Action indicators e.g. proportion of properties at risk benefiting from flood defences and extent of impermeable surfaces in urban areas.

Indicators for Increasing damp and mould in buildings (Data sources: Scottish House Condition Survey).

- Extent of the damp etc.
- Resulting Actions: New policies on standards for rented accommodation etc.

Indicators for Energy Demand: (Data sources e.g. Official Energy Efficiency Ratings; General Register Office Scotland; Scottish House Condition Survey)

- Opportunity: Milder winters will reduce energy demand for heating.
- Risk indicators e.g. property energy efficiency and average household size
- Action indicators: proxies: extent of insulation measures in housing; extent of energy monitoring by occupants

Indicators for Transport and Flooding: (Data sources e.g. SEPA, Transport Scotland/local authorities, Scottish Flood Defence Asset Database; Network Rail; Scot Rail)

- Risk of flooding of roads, disruption due to flooding of roads, flooding of railways
- Risk indicators: e.g. length of road network affected, rural road links at risk
- Impact indicators: e.g. number of flood events, direct cost of physical damage
- Action indicators: e.g. length of road network at risk benefiting from flood protection

Indicators for Transport and Landslides (Data sources include SEPA floods/flows; Transport Scotland and Local Authorities).

- Risk Indicators for landslides affecting roads; scour of road and rail bridges
- Risk indicator: no data on landslide risk therefore a proxy used eg. bridges not designed or modified to account for scour.
- Impact indicator e.g. Roads closed due to landslides; failures or unplanned closures due to scour.
- Action indicators e.g. Spend on landslide mitigation measures and spend on bridges where remedial action taken to reduce scour
- Indicators for Transport Dependency:
- Risk that key workers are unable to get to work.
- Action indicator: extent of travel plans of relevant organisations which account for key workers being able to get to work in extreme weather/cases of infrastructure failure.

7 Challenges and Lessons

In developing the Scottish Climate Adaptation Indicators many pitfalls have been encountered. For example, it is hard to get indicators which are specific just to climate change because of the complexity of the systems involved (e.g. there may be changes in agricultural productivity due to climate but also due to land management practices and policy support). Most of the things we are trying to monitor are part of complex systems. For example energy demand for heating can be influenced by factors that range from outdoor temperature (the only climate related part), to household size (ie more people choosing to live alone), or whether people are willing to wear thick jumpers in winter at home. There is effectively no data available for some of the most important factors (chosen indoor temperature, extent of home heated, time periods of heating chosen).

It has proven to be best to try to understand the relative contribution of climate change amongst a set of wider drivers of change, rather than trying to specifically allocate certain quantities of change to adaptation. It is better to understand whether the direction of travel is right, and whether, given the combination of factors at play, there are useful adjustments that can be made to the quantity or type of adaptation actions. For instance the mind-set around the problems of 'Overheating in Homes' in Scotland is currently such that people do not think it a possibility here and yet there is growing evidence that in many current modern homes that are poorly designed climatically it is already a significant problem despite the fact that there have been few heat waves in the region in recent years. This begs the question as to whether we should adopt a more global set of indicators where regional shifts in impacts are premeditated.

To maximise their effectiveness the Scottish team has chosen to develop abroad suite of indicators which frame and assess adaptation progress and resilience to climate change over time, rather than develop objective indicators which measure every component of an intervention as might be applicable with more straight forward issues. Climate change demands a broader set of indictors than is often the case so we are using a combination of:

(i) High level cross cutting indicators measuring resilience – things around which there is consensus are 'good things' to do for a variety of purposes (win-win actions – for example, implement policies that will improve the level of insulation and state of repair in households).
(ii) More specific indicators where broad indicators are not so useful or to provide additional insight into important adaptation issues (eg percentage of homes with significant penetrating damp).

To ensure indicators are relevant and useful we have engaged stakeholders throughout to achieve support for the indicators; ensure the relevance of indicator metrics to decision makers and making; access to data and fit with existing monitoring and to maximise their impacts and the use of their outputs.

There are many data challenges in developing robust and durable indicator sets and there may be tensions between modeled and real data. In flooding it is important to distinguish between data derived from modeling and data derived from actual events. Additional data sets about actual flood events and flood damage must be developed in order to track adaptation to increased flood risk in a meaningful way and this implies cost and effort and political commitment to make that happen.

Often one cannot use the 'ideal' indicator, simply as no data is available for it, or the data quality and continuity is not good enough.

It is important to be explicit about indicator and data quality:

- To reflect the policy and conceptual relevance of the indicator
- Its sensitivity and specificity to climate change and the issue in question

- To reflect the quality of the underpinning data e.g. frequency of collection, time span over which data is available, spatial scale, resolution of data and so on

Issues around this include:

- Lack of appropriate data which results in less than ideal indicators being used;
- Inclusion of indicators which do not have a strong climate signal;
- Trying to get agreement from data owners to take responsibility for updating indicator data/updating the actual indicators

In developing the Scottish Adaptation Indicator sets we concluded that, where possible, it is best to:

(a) Use existing indicators and data to avoid duplication
(b) Design the indicators to ensure rapid absorption into the main-stream
(c) Make efforts to get 'voter traction' and interest
(d) Balance suitability and data availability
(e) Make trade-offs where necessary
(f) Be explicit about the minimum quality of data acceptable for the indicators
(g) Leave time and funds to pursue data as access to data can involve negotiations
(h) Secure licence agreements
(i) Not underestimate the possible future spend necessary to fund data gaps
(j) Design the data set with a clear plan for how it will updated as new data becomes available.

8 Conclusions

Adaptation monitoring is particularly challenging as the impacts of adaptive actions are often multi-faceted and not always short-term and many occur over many years. Adaptation planning needs to be flexible to meet the needs of changing circumstances involving highly variable rates of change in different facets of the climate itself. They must apply within the context of locally and regionally evolving social and economic circumstances and in light also of new knowledge. Adaptation cannot be seen just as an end point but as a rapidly evolving phenomenon and indicators of it too must be capable of responding to those changes.

Challenges have been many in the Scottish experience. For us this has been pioneering research in developing our adaption indicator sets. But on talking to similar groups around the world we are all too often locked into our own pioneering trajectories. We ask common questions such as 'what else needs to be done to ensure that indicators are used to increase resilience and adaptation?' In sharing our experiences and lessons learnt in Scotland we hope to contribute to the developing discussions of the challenges involved with a view to developing useful and compatible datasets and indicators that can contribute to the adoption and implementation of more successful adaptation indicators not only in Scotland but in many cases across the world.

Glossary of Scottish Government Adaption Indicator Framework Terms

Adaptation Actions: Responses which exploit the opportunities of climate change or adjustments aimed at helping society and environment cope with the risks of a changing climate.

Exposure: The level of climate stress upon the system (e.g. upon the natural environment, built environment or on people).

Impact: The felt consequences of climate change. These may be economic, structural, social, or environmental.

Risk and opportunity: The potential impacts of climate change assessed according to their likelihood, magnitude of potential impact and urgency.

Vulnerability: The degree to which a system is susceptible to, or unable to cope with, adverse effects of climate change

References

Adaptation Sub-Committee (2011) How well is Scotland Preparing for Climate Change. London, Climate Change Committee: www.theccc.org.uk/publication/how-well-is-scotland-preparing-for-climate-change-asc-scotland-report-2011/

Berrang-Ford L, Ford JD et al (2011) Are we adapting to climate change? Glob Environ Chang 21(1):25–33

Crichton D (2001) A Scottish lead in managing flood isk. Town Country Plan 6:188–189. (See also Roaf et al. 2009)

DEFRA (2004) Review of UK climate change indicators. DEFRA, London

DEFRA (2012) UK climate change risk assessment: Government report. www.gov.uk/government/uploads/system/uploads/attachment_data/file/69487/pb13698-climate-risk-assessment.pdf

Engle NL (2011) Adaptive capacity and its assessment. Glob Environ Chang 21(2):647–656

Eriksen S, Kelly PM (2007) Developing credible vulnerability indicators for climate adaptation policy assessment. Mitig Adapt Strateg Glob Chang 12(4):495–524

Hofmann ME, Hinkel J et al (2011) Classifying knowledge on climate change impacts, adaptation, and vulnerability in Europe for informing adaptation research and decision-making: a conceptual meta-analysis. Glob Environ Chang 21(3):1106–1116

IPPC (2001) Climate change 2001: impacts, adaptation, and vulnerability: contribution of Working Group II to the third assessment report of the Intergovernmental Panel on Climate Change. Cambridge, Cambridge University Press

IPPC (2007a) Climate change 2007 impacts, adaptation and vulnerability. In: Parry JP, Palutikof PJ, van der Linden and Hanson CE (eds) Working Group II contribution to the fourth assessment report of the IPCC. O. F. C. M.L. Cambridge University Press, Cambridge, UK

IPPC (2007b) Definitions of key terms within Climate Change 2007: Working Group II: impacts, adaptation and vulnerability. Retrieved 8 Nov 2011, from http://www.ipcc.ch/publications_and_data/ar4/wg2/en/frontmattersg.html

Levina E, Tirpak D (2006) Adaptation to climate change: key terms. Organization for Economic Co-operation and Development, International Energy Agency, Paris

Metzger MJ, Rounsevell MDA et al (2006) The vulnerability of ecosystem services to land use change, *Agriculture*. Ecol Environ 114(1):69–85

Natural England (2010) Climate change adaptation indicators for the natural environment. N. Williams, Peterborough, Atkins, p 68

Roaf S, (2013) Transitioning to eco-cities: reducing carbon emissions while improving urban welfare, Chapter 7. In: Young-Doo Wang, John Byrne (eds) Secure & green energy economies. Transaction Publishers, Washington, DC

Roaf S, Crichton D, Nicol F (2009) Adapting buildings and cities for climate change. Architectural Press, Oxford

Scottish Government (2014) Climate ready Scotland: Scottish climate change adaptation programme. Retrieved 6 June 2016 at: www.gov.scot/Resource/0045/00451392.pdf

Tyndall Centre (2004) New indicators of vulnerability and adaptive capacity. Accessed on August 16, 2016 at: http://www.tyndall.ac.uk/sites/default/files/it1_11.pdf

Chapter 5
The Sustainable Portion of Gross Domestic Product: A Proposed Social Ecological Economic Indicator for Sustainable Economic Development

Abbas Mahravan and Brenda Vale

Abstract This article discusses a way of measuring the sustainable portion of Gross Domestic Product (GDPs) and argues for its use as an ecological-social economic indicator to evaluate the economic footprint of human activities and products as being environmentally and culturally sustainable. It proposes a method for calculating GDPs based on four quantitative factors: the domestic product or economic activity, its ecological footprint, its overshoot portion of energy use, and the cost of generating the overshoot portion of energy use using renewable resources (in this study wind and solar electric systems). As a case study of using GDPs as an indicator, this article investigates the economic footprint of the Otago Central Rail Trail (OCRT) which has been cited by the New Zealand Tourism Strategy 2015 as a successful tourism project. Using GDPs the finding for the OCRT is that 43.6 % of total OCRT GDP must be spent to restore the environment damaged through its development.

Keywords Sustainability • Ecological economic indicator • Ecological footprint • GDPs

1 Introduction

The aim of developing a way of measuring the sustainable portion of Gross Domestic Product (called here GDPs) as a new indicator for sustainability was to be able to measure the impact of increased economic growth, with the further aim of finding

---

A. Mahravan (✉)
Department of Architecture, Razi University, Kermanshah, Iran
e-mail: mahravan_abbas@yahoo.com

B. Vale
School of Architecture and Design, Victoria University of Wellington, Wellington, New Zealand

T.H. Karyono et al. (eds.), *Sustainable Building and Built Environments to Mitigate Climate Change in the Tropics*, DOI 10.1007/978-3-319-49601-6_5

ways in which improved economic performance could have a neutral or even a reduced impact on the environment.

The first need is to establish what is meant by sustainability. Here, sustainability is seen as a key factor in defining appropriate products and activities and in determining the main policies and strategies for lasting socio-economic development. The reasons for this are discussed below.

Appleby (2011) argues that the roots of modern sustainability and related energy policy lie in the scenario proposed by the 18th century political economist Thomas Malthus, commonly known as the Malthus Catastrophe or Nightmare. In this scenario Malthus forecasted that population would grow faster than agricultural production (Appleby 2011). Since the onset of the industrial revolution in the latter part of the eighteenth century and the subsequent development of capitalism, numerous catastrophic events have happened in the world (for instance famines and conflict in African countries such as Ethiopia and the Sudan) that suggest the truth of the 'Malthus Catastrophe or Nightmare' as people struggle for control of insufficient resources.

The World Commission on Environment and Development (WCED) has put forth a definition of sustainability as "meeting the needs of the present without compromising the ability of future generations to meet their needs" (WCED 1987). This concept of sustainable development, established in the Brundtland Report (WCED 1987), attempts to support continuous development of human societies for the present and future generations. The concept appears to be based on the idea of balance between growth in social, economic and environmental aspects, which together form the central tenets for development of a sustainable society (Robin and Poon 2009). However, this very definition raises questions, as 'growth' in environmental aspects within a finite system like planet earth is itself not a sustainable activity.

2 Sustainable Development Goals

Sustainability is an inherently dynamic, indefinite and contested concept. It is not a goal to be achieved, but rather the establishment of a system of learning and adaptation, recognising that there are limits to the resources the planet can supply (Mog 2004). This suggests that flexibility and adaptability in terms of different cultural, environmental and economic situations are factors that could be used as frameworks to evaluate sustainability.

To achieve sustainable development, indigenous people and their needs, their natural resources, and the development of conservation policies to protect them, together with their local economic systems and shift toward sustainability, should all be viewed as the main components of progressive strategies. If a development project ignores some of these components during its programming, management and implementation, it cannot be called a sustainable project. For example, Mog (2004) argues that a project that reduces land degradation but at the same time increases poverty or inequality, or achieves its goals through the establishment of rigid institutions (e.g. forcing all farmers off a parcel of land and establishing a strictly exclusionary protected area) would not be considered successful in promoting sustainable development, no matter how great the improvement in environmental quality. On the other hand a project which enhances environmental quality in a culturally sensitive manner (e.g. through extension of appropriate conservation farming techniques) without having any positive impact on poverty, is also grossly insufficient. As a result, projects which create significant positive change in many realms while generating little or no negative impacts can be viewed as sustainable activities. As a result, the following part of this article takes the environmental, social and economic outcomes of sustainability as the main components of a framework for sustainable development.

2.1 Environment

One of the main aims for sustainable development is to reduce use of non-renewable energy resources. Nowadays, humans are faced with issues such as the uncontrolled growth of population, dramatically expanding urban areas and, consequently, increasing needs to use more energy resources, as well as degradation of the environmental heritage and unequal distribution of capital. Because of industrialisation and the expansion of modernisation, urban areas throughout the world have increased in size and population. According to Omer (writing in 2008), about 50 % of the world's population were by then urban dwellers. In addition, he states that all population growth expected between 2000 and 2030 will be concentrated in urban areas because of the rural-to-urban population shift. Furthermore, with an expected annual growth of 1.8 %, the world's urban population will double in 38 years. In terms of resource consumption, at present 2 % of the land surfaces in the world are covered by cities, yet the people living in them consume 75 % of the resources consumed by humankind. Omer (2008) points out that the ecological footprint of cities is many times larger than the areas they physically occupy. Continued growth of the world's population has caused serious need to use all sources of energy, particularly fossil fuels like oil and gas. Moreover, as researchers such as Appleby (2011) show, if everyone in the world lived like the average North American five planets would be needed rather than the one available, which is obviously unsustainable (Appleby 2011).

The 2002 World Summit on Sustainable Development in Johannesburg committed itself to supporting and encouraging the expansion of renewable energy resources to accelerate the shift towards sustainable consumption and production. It could be argued that the main goal for the summit was to make productivity independent from resource use. According to Omer (2008), breaking the linkage between resource use and productivity can be achieved by trying to ensure economic growth does not cause environmental pollution. Furthermore, he goes on to suggest the main policies for achieving the goal of a pristine environment are; improving resource efficiency; examining the whole life cycle of the product; enabling consumers to receive more information on products and services; and examining taxes, voluntary agreements, subsidies, regulation and information campaigns regarding the best stimulation to innovation. However, this approach still fails to recognise the fundamental problem that there must be a limit to growth within a finite system.

2.2 *Society*

One of the basic strategies to achieve the goals associated with sustainability is to make societies aware of the components and structure of sustainability, its policies and benefits. In 1991, the World Conservation Strategy (cited in Robin and Poon 2009) stated that the principles of a sustainable society include; respect and care for the community; improvement in the quality of human life; conserving the earth's vitality and diversity; minimising the depletion of non-renewable resources; keeping within the earth's carrying capacity; changing attitudes and practices; enabling communities to care for their own environments; providing a national framework for integrating development and conservation; and creating a global alliance. According to the latter principles, change of attitudes and practices can be viewed as the focal point for a cultural shift toward sustainability. Robin and Poon (2009) point out that change of attitudes and practices means a transformation of mind-set and actions. These criteria demand changes in the innate basic assumptions, values, beliefs and behaviour that are present in the established culture of society.

2.3 *Economy*

Graff et al. (1996) show that in sustainable development, economic, cultural and social developments are interrelated. They further argue that in sustainable development environmental problems cannot be isolated from others such as poverty and social disintegration. Likewise they argue that economic growth makes poverty alleviation easier and cultural development can change the pattern of the consumption of natural resources in order to ensure their protection. However, the main outcomes of unsustainable economic growth arising from capitalist-based development can be considered as making poor people poorer and rich people richer.

Vera and Langlois (2007) apart from believing energy to be vital for eradicating poverty, point out that about one-third of the world's population still relies on the use of animal power and non-commercial fuels, with about 1.7 billion people lacking access to electricity. They also suggest that socio-economic development as an integral part of sustainable development is severely limited by the lack of access to modern energy services. According to the World Bank (2012) in 2008, 1.29 billion people lived on an income below US$1.25 a day, equivalent to 22.4 % of the population in the developing world. Although global poverty has dropped from 43 % in 1990 and a higher 52 % in 1981, one billion people still experience critical poverty (World Bank 2012). According to the World Bank Development Indicators (2008), in 2005 the majority of all resources (76.6 %) were consumed by 20 % of the world's richest people, 21.9 % of all resources by 60 % of the world's middle-income people and only 1.5 % by the 20 % that make up the world's poorest people.

3 Sustainability Indicators

As discussed above, sustainable development is generally viewed as a multi-complex phenomenon that rests on the three interrelated pillars of economic growth, social progress and protection of the natural environment. Since all dimensions of sustainable development are interrelated, its economic footprint cannot be evaluated as being sustainable without the existence of a comprehensive framework so that economic indicators cover all the environmental and social footprint of any economic development. Although different sustainable development frameworks and indicators have already been suggested, as many researchers and related organisations such as NZCTU (2010) and UNEP (2011) have shown, the use of conventional economic indicators, such as GDP and other macroeconomic aggregates, can lead to a distorted picture of economic performance, particularly since such measures do not reflect the extent to which production and consumption activities may be drawing down natural and social capital. Before introducing the proposed indicator of GDPs, a brief introduction to sustainable economic indicators, their strengths and weaknesses, and some examples, are given below.

3.1 Economic Indicators

Golusin and Ivanovic (2009) argue that there are differences between the indicators of sustainability and the traditional economic, social and ecological indicators, such as profit of interest groups or stakeholders, percentage of inhabitants with an asthmatic condition, and water quality (as respective examples of economic, social and ecological indicators).Taking this further, Spangenberg et al. (2002) point out the necessary key qualities of suitable sustainability indicators, explained as their guiding function for political decision making towards sustainable development. They show that sustainability indicators ought to structure the basis for improved information and data collection, and enable comparative and country specific analysis of the state of and progress towards sustainable development. This article would further suggest such indicators should also refer to the anticipated environmental, social and economic outcomes, and be able to address the success in or lack of achieving these outcomes. In addition, as Nader et al. (2008) explain, sustainable indicators should provide hard, quantitative data to ensure a sound basis for all environmental, social, and economic policies for the future. However many researchers such as Barrass et al. (1997), Rennings and Wiggering (1997), Spangenberg et al. (2002), Lawn (2003), Nourry (2008) and Bagstad and Shammin (2012) determine that most proposed economic and socio-economic indicators fail to reflect both holistically and comprehensively the linkage between economic growth, environmental conservation and social wellbeing and development in a sustainable way.

This article views the shortage of efficient economic indicators as one of the critical issues for measurement of the economic footprint of an activity (such as ecotourism) or a product (such as architecture) as being economically sustainable. Many attempts have been conducted to introduce economic indicators that have the ability to cover the social and ecological dimensions of sustainable economic development but the results contain potential weaknesses that must be justified. For instance, the NZCTU (2010) proposes taxation as a strategy to monitor and control the environmental degradation caused by economic development. They go on to suggest that polluters should face taxes on their emissions, including greenhouse gas emissions, with the aim of them paying the costs of all significant 'externalities' (side effects such as pollution and global warming), but how the level of environmental pollution or degradation can be measured is a question still to be answered. There are also weaknesses in the current economic indicators of Gross Domestic Product (GDP) and the Genuine Progress Indicator (GPI) as measures of moving towards sustainability, as discussed below.

3.1.1 Gross Domestic Product (GDP)

GDP is a very common macroeconomic indicator, having been used in the United States since 1963 (Bagstad and Shammin 2012). It is a measure of the monetary value of the annual production of goods and services in a domestic location, so covers the natural and human created capital in a particular place, such as a nation. It can be calculated in nominal or real values (Lawn 2003). Lawn further argues that if GDP is measured in nominal values, it is measured in terms of the prices at the time of production and if in real values, in terms of the prices of all goods in a particular year, which is often referred to as the base year. Consequently, differences in the quantity of goods and services produced from year to year can be reflected by annual changes in real GDP.

In 1992, the United Nations Conference on Environment and Development in Rio de Janeiro underlined the limitations of gross domestic product (GDP) as a measure of sustainable development for a country. Agenda21 (1992) found GDP to be an example of common indicators that do not assess the sustainability of an economic system. Rennings and Wiggering (1997) suggest that measuring the economic and environmental aspects of sustainable development need to be complementary, while Barrass et al. (1997) go further by stating that growth in GDP will only align with sustainable development if the environment and the resources it supplies are valued in the same way. The same authors go on to say that conventional ways of assessing GDP tend to undervalue the vital natural resources supplied by the environment, since no account is taken of their replacement times. Bagstad and Shammin

(2012) identify five common criticisms of GDP: (1) it consists of simple addition of all expenditure, even expenditure that does not improve social welfare, such as that spent on defence; (2) it excludes the value of unpaid household labour and volunteer work in the community; (3) it fails to account for the value of non-renewable resource depletion while total resource stocks decline over time; (4) it fails to account for income distribution, poverty, and the costs of inequality; (5) it lacks inclusion of the positive contributions from natural, human and social capital types.

3.1.2 Adjustment to GDP

Given the shortcomings of GDP, economists have proposed adjustments to it to make it a better measure of social wellbeing. One of the notable examples is Daly and Cobb's (1989) Index of Sustainable Economic Welfare (ISEW) which was later revised as the Genuine Progress Indicator (GPI). As Mahravan (2012) reveals ISEW/GPI is a composite indicator which includes economic, social and environmental indicators. Examples of the former are "Personal consumption per capita" and "Value of household labour", social indicators include "Cost of crime" and "Loss of leisure time" and environmental indicators are based on measurables such as "Cost of air pollution" and "Loss of forest cover"(Bagstad and Shammin 2012). GPI calculation includes the cost of environmental degradation as well as factors for natural capital regeneration but is still somewhat arbitrary in what is included and excluded (Pulselli et al. 2006). In addition Pulselli et al argue that since ISEW/GPI is composed of a list of items which indicate economic welfare or sustainability ISEW/GPI cannot be used as an indicator of the real level of economic welfare at the same time as being an indicator of sustainability. As explained by Pulselli et al. (2006) ISEW/GPI does not really achieve its anticipated outcomes as it was based on the strong sustainability model within which it is not possible to substitute human capital for natural capital. Furthermore the authors go on to say they believe ISEW/GPI does not make a clear distinction between national and natural capitals or between the two forms of natural resources (non-renewable and renewable). Nourry (2008) also points out that in ISEW/GPI, the incorporation of the cost of environmental degradation is not sufficient to determine sustainability or otherwise, so it cannot be strictly considered as an indicator of sustainable economic welfare.

4 Sustainable Portion of GDP (GDPs)

This study aims to help develop economic indicators that have the ability to calculate economic progress and that can also be comprehensive measures of well-being and environmental sustainability. To achieve this aim, an economic indicator is needed that is as clear and appealing as GDP but more inclusive of other dimensions of progress, in particular the environmental and social aspects. As proposed by Mahravan and Vale (2011), the development of local domestic products can be viewed as a fundamental social-cultural priority in a sustainable development strategy. Given this, progress can clearly be measured economically using GDP, but to measure the environmental impacts of these local products, it is essential to integrate both gross domestic product and the cost that must be spent to conserve or restore the environment which is damaged during the process of development. GDPs is the proposed indicator for doing this and is explained further below.

4.1 GDPs Equation

This research proposes a method to calculate the sustainable portion of GDP (called here GDPs), and hence reveal the non-sustainable portion of GDP. GDPs can be used as an economic indicator which measures both the social-economic benefits of producing local domestic products and their environmental impacts in an integrated concept. The method to measure GDPs involves five factors which are listed below, and then explained in detail:

(a) EF (gha – see definition below): Ecological footprint of an activity or a product.
(b) GDP ($): Earned GDP from this activity or product.
(c) EF1 (gha): Overshoot portion of EF of the product or activity relative to the available bio- capacity.
(d) E1 (GJ): Overshoot portion of the life cycle energy use of the product or activity relevant to the energy equivalent of bio-capacity.
(e) CE1 ($): The cost to generate the overshoot portion of the life cycle energy through the use of renewable resources.

4.1.1 Ecological Footprint (EF)

The Ecological Footprint (EF) measures the extent to which humanity is using nature's resources faster than they can regenerate. EFs are usually presented together with bio-capacities (BCs), which measure the available bio-productive land area. For the calculation of GDPs, the EF of a given product or an activity is equal to the area (in global hectares of average productivity – gha) needed to absorb the CO_2 emissions generated through using fossil fuel to produce its life cycle energy use (GJ).

(a) **EF1**

If an EF is larger than the available BC for a selected time period, the EF/BC resource accounting produces a deficit or overshoot EF (EF1). A deficit occurs in the case where human resource extraction and waste generation exceed an ecosystem's ability to regenerate the extracted resources and to absorb the generated waste.

(b) **E1 (GJ)**

This is the overshoot portion of the life cycle energy use of the product or activity such that its CO_2 emissions are more than the related BC to absorb them.

(c) **CE1 ($)**

CE1 is assumed equal to the money needed to generate the overshoot portion of the life cycle energy used (E1) through the use of renewable rather than non-renewable resources. Its value depends on the level, type and cost of the available technology and renewable resources. As a result, CE1 can be different from one location to another, or from region to region, or country to country. The following equation can be used to evaluate the sustainable portion of GDP (GDPs):

$$\text{GDPs} = GDP - CE1$$

4.2 *The Main Characteristics of GDPs*

GDPs is introduced here as a multi-dimensional indicator that has the ability to measure and evaluate the socio-economic and environmental success of human activities and human made products as being sustainable.

4.2.1 Social-cultural Characteristics

Making opportunities for local and indigenous people to produce and present their cultural capitals and products is one of the fundamental cultural outcomes for the sustainable development of an activity such as ecotourism. Since GDPs emphasises the development of local and domestic products, it can be considered as a social-economic indicator that contributes to the evaluation of success in achieving this cultural outcome.

4.2.2 Environmental Characteristics

GDPs is an economic indicator that is sensitive to the environmental impacts of human products and activities. Integration of EF and economic development in GDPs allows it to be considered as a social-economic indicator as well as an ecological indicator. The difference between GDP and GDPs (GDP- GDPs) addresses the amount of GDP that must be spent to conserve and restore damaged environmental resources. It can contribute to the more accurate accounting of other economic indicators.

5 Case Study: Otago Central Rail Trail (OCRT)

5.1 Otago Central Rail Trail (OCRT) – Background

Originally the Otago Central Railway, located in the South Island of New Zealand, was constructed to transport gold from the booming Central Otago goldfields of the late 1800s and into the busy city of Dunedin (Graham 2009). In the forty years it took to complete the railway, over 60 bridges and 3 tunnels were built between Middlemarch and Clyde. According to Burns and Corbet (2007) the railway closed in 1990, at a time when the rural community was in a state of stagnation. Graham (2009) points out that through the early 1990s, towns in Central Otago as in other rural areas in New Zealand lost many of their services and suffered from a reducing population, as banks and post offices closed and many businesses moved away. However, some local people viewed the closed railway line as an opportunity and potential attraction for developing the tourism industry in Central Otago.

The Department of Conservation (DOC), which had bought the land in 1993, was persuaded to consider the potential for outdoor recreation that the rail land offered. DOC spent six years and more than $850,000 preparing, upgrading and restoring the closed railway and related historical heritage and infrastructure to form a walking and cycling route, now viewed as one of the main attractions for Otago Central Rail Trail (OCRT) visitors (Burns and Corbet 2007). The OCRT was established in February 2000 through a cooperative effort between DOC and the Otago Central Rail Trail Trust, which had been formed in 1994 (Jellum and Reis, 2008). Since that time, the positive economic and environmental impacts of the OCRT have changed local attitudes and created greater community pride because of its cultural impacts, turning those who were originally against its creation into tourism service providers through farm stays, farm visits, cafes, and accommodation (Burns and Corbet 2007).

The 150 km recreational rail trail runs west from Middlemarch, travels along the Taieri River valley and through the Maniototo Plains, crosses the Rock and Pillar, North Rough, and Raggedy Ridge ranges, and passes through Ida Valley and Poolburn Gorge along the Manuherikia River (OCRT official web site). A trip along the entire 150 km length of the Otago Central Rail Trail may begin at either end, at Clyde or Middlemarch, and can be cycled in approximately 4 days, walked in approximately 6 days, or experienced by any number of shorter station to station trips. The OCRT's highest point is near the township of Wedderburn, 618 m above sea level, with gentle descents in either direction. Whether visitors are engaged in a multi-day excursion or a one day trip, a large selection of accommodation providers, restaurants, and other businesses support these rail trail visitors both in communities adjacent to the rail trail and at nearby off-trail locations.

Table 5.1 Total EF of OCRT (2011)

Category	EF (gha)	% of total footprint		EF (gha/visitor)
Transportation	1167	72.2		0.1
		Domestic	International	
		32	38.2	
Food	401.2	24.8		0.034
Accommodation	42.4	2.6		0.0036
Activities	6.57	0.4		0.00056
Total	1617	100		0.138
Total OCRT visitor numbers = 11,788				

5.2 *OCRT: Ecological Footprint*

Table 5.1 demonstrates that the 2011 EF of the OCRT is equivalent to 1617 gha, which derives from the integration of the calculated EFs of transportation, food, accommodation services and activities used and engaged in by 11,788 OCRT visitors that year (Mahravan 2012). As shown in Table 5.1 transportation has the largest footprint of 1167gha (72.2 %) of the total EF, followed by 401.2 gha (24.8 %) food, 42.24 gha (2.6 %) accommodation services and 6.57 gha (0.4 %) visitor activities. The total EF of transportation can be further divided into 38.2 % for international transportation and 34 % for domestic transportation (Table 5.1). This shows the importance of getting to a tourist destination in its overall environmental impact.

5.3 *The EF of Sustainable Living*

Vale and Vale (2009), using data from the Cardiff Report (2001), propose the EFs of sustainable living for the ten categories of life activities and products shown in Table 5.2 and compare these proposed EFs with their present amounts. Vale and Vale (2009) classify the ten categories into the two areas of individual and collective products and activities (Table 5.2). The individual products and activities of food and drink, domestic energy, domestic travel, "stuff" (consumer goods), housing, and holidays, need to have EFs in the range of 0.03–0.43 gha/capita in order to be considered as being sustainable (Table 5.2). The sustainable EFs of the collective categories, including infrastructure, government, and services, are calculated to be in the range of 0.24–0.47 gha/capita. The total EF of sustainable living (as the goal) is equivalent to 1.85 gha/capita, (this represents the available per capita footprint if the available bio-productive land were divided between the global population) and the total EF for when the evaluation was done is 6.41 gha/capita.

As presented in Table 5.2, the sustainable EF of holidays is calculated to be 0.03 gha/capita and, in the Cardiff report (2001), this includes the EFs of transportation, food, accommodation and waste water that are consumed/used/produced by Cardiff visitors. This study uses this sustainable EF of holidays as the goal to be achieved through the sustainable development of the OCRT and compares its present

Table 5.2 The ecological footprint of sustainable living (Vale and Vale 2009)

	Now	Goal
Individual		
Food and drink	1.33	0.43
Domestic energy	0.99	0.33
Domestic travel	0.99	0.33
Stuff	0.64	0.21
Housing	0.16	0.05
Holidays	0.10	0.03
Individual total	5.00	1.38
Collective		
Infrastructure	0.74	0.24
Government	0.41	0.14
Services	0.26	0.09
Collective total	1.41	0.47
Total revised footprint	6.41	1.85

Table 5.3 The ecological footprint of OCRT- sustainable tourism (2011)

Overshoot portion of OCRT visitors- EF (gha/visitor/year)		
Now	Goal	Overshoot EF(gha/visitor/year)
0.138	0.03	0.108
Total overshoot EF (gha) of 11,788 OCRT visitors = 1273		

EF with its target EF to calculate the overshoot portion of the current EF. Table 5.3 compares the present EF (0.138 gha/visitor) of the OCRT (see Table 5.1) and its goal (0.03 gha/visitor) for being environmentally sustainable. The difference between these two EFs (0.138 gha/visitor – 0.03 gha/visitor) gives the overshoot portion of the OCRT EF, which is 0.108 gha/visitor (Table 5.3).

5.4 *The Overshoot Portion of OCRT EF- Energy*

As EF (gha) = energy used (GJ) ÷100 (GJ/gha) (global average carrying capacity of land relative to CO_2 absorption) (Wackernagel and Rees 1996);

Consequently:

Energy used (GJ) = EF (gha) × 100 (GJ/gha)

As shown in Table 5.1 the EF of OCRT is 0.138, so its energy use is equivalent to:

0.138 (gha/visitor) × 100 (GJ/gha) = 13.8 (GJ/visitor)

Moreover its sustainable energy use is:

0.03 (gha/visitor) × 100 (GJ/gha) = 3.0 (GJ/visitor)

The overshoot portion of OCRT energy use is:

13.8 GJ/visitor – 3.0 GJ/visitor = 10.8 GJ/visitor

As a result the overshoot energy use of 11,788 OCRT visitors is equivalent to:

10.8 (overshoot energy use GJ/visitor/year) × 11,788 (visitors) = 127,310.4 GJ/year

5.5 *OCRT: Economic Gross (GDP)*

According to Central Otago District Council (CODC) (2011) in 2011, the direct output of the OCRT was equivalent to NZ$8,159,042 and its total output was $12,277,024 (Table 5.4). Likewise, in 2011 the OCRT direct contribution to GDP was $4,213,793 and its total contribution was $6,245,289. Table 5.4 also shows that in 2011, 121 full time staff equivalents were in employment related to the OCRT.

In 2011 of the total $6,245,289 OCRT contribution to GDP, accommodation formed the majority of $2,837,412, followed by vehicles and equipment hire ($975,119), and business administration and managerial services ($900,602) (Table 5.5). Other categories included food and consumables, road passenger transport, retail, and entertainment, with GDP contributions in the range of $61,925–$782,795 (Table 5.5).

5.6 *The Cost of Generating Energy Through Using Renewable Energy Sources*

This study proposes two scenarios to calculate the cost of generating energy from renewable resources. In the first the cost (wind + solar systems) is calculated based on the anticipated reduction in cost of solar electric systems (PVs) by a factor of 10 from current prices. In the second scenario the total cost is calculated based on the current PV costs. The reason for doing this will become clear from the results.

Table 5.4 The economic impact of the OCRT on the Central Otago/Otago economy (Central Otago District Council (CODC) 2011)

Impact summary	Direct	Total
Output	$8,159,042	$12,277,024
GDP	$4,213,793	$6,245,289
Employment (FTE's)	96.0	121

Table 5.5 Economic impact of OCRT- by categories. (CODC 2011)

Category	Total GDP
Accommodation	$2,837,412
Food and consumables	$782,795
Retail	$241,299
Road passenger transport	$446,136
Vehicle and equipment hire	$975,119
Business administration and managerial services	$900,602
Entertainment	$61,925
Total	$6,245,289

5.6.1 First Scenario

The average cost of wind and solar technology in New Zealand (Table 5.6) is used to calculate the cost of generating 1GJ of renewable energy. The figure for solar is based on the cost of $350-$500 per MWh in New Zealand given by EECA (2012) but reduced by a factor of 10 based on decreasing PV cost data from the National Renewable Energy Laboratory in the United States (US Department of Energy 2011).The average cost of using wind and solar systems in New Zealand is assumed here to be NZ$71.25/MWh or NZ$19.8/ GJ.

As shown in Table 5.7, in 2011 the total OCRT GDP was $6,245,289 and its overshoot energy use is equivalent to 127,310.4 GJ/year. Since the average cost of generating 1GJ of energy through using wind and solar systems is NZ$19.80 the consequent cost of generating the total OCRT overshoot energy use is equivalent to (127,310.4 GJ/year × NZ$19.80) = NZ$2,720,746. The OCRT GDPs is equivalent to its total GDP ($6,245,289) less the total cost to generate the overshoot portion of its energy use by wind-solar systems (NZ$2,720,746) = NZ$3,524,543 (Table 5.7).

Figure 5.1 makes a comparison between the OCRT's GDP and its GDPs. As illustrated OCRT GDPs is NZ$ 2,720746 (43.6 %) less than OCRT GDP (NZ$6,245,289). On the other hand, 43.6 % of total OCRT GDP must be spent to reduce its environmental impacts to the level of sustainable living as related to the

Table 5.6 New Zealand energy costs

Energy technology	Cost (NZ$/MWh)	Average cost (NZ$/MWh)[a]
Wind	80–120[b]	100
Solar system	35–50[c]	42.5
Average cost of energy generated by wind-solar systems (NZ$/MWh)		71.25

[a]NZ$1= US$0.83 (http://themoneyconverter.com) (Viewed on February 2011)
[b]Reference: New Zealand Wind Energy Association, 2011. (http://windenergy.org.nz/wind-energy/costs)
[c]Reference: www.Energywise.govt.nz

Table 5.7 OCRT-sustainable portion of total GDP (GDPs)

Total OCRT GDP (NZ$)	6,245,289
EF (gha)	1617
Overshoot EF (gha)	1273
Overshoot energy use (GJ)	127,310.4
Energy cost (NZ$/GJ) wind-solar	19.80
Total cost (NZ$) to generate the overshoot portion of OCRT energy use by wind-solar systems	2,720,746
GDPs (NZ$)[a]	3,524,543

[a]GDPs = Total GDP – Total cost to generate the overshoot portion of energy use through the use of renewable resources

EF of holidays (Table 5.2). As explained above, this calculation of GDPs is based on the assumption of a reduced cost for solar energy following historic cost trends.

5.6.2 Second Scenario

In the second scenario (Table 5.8) the average cost to generate 1GJ of energy using renewable resources (wind + solar systems) is equivalent to NZ$ 262.5/MWh. Since 1 MWh = 3.6 GJ, in the second scenario the cost to generate 1GJ of energy using renewable resources is NZ$ 72.90 (262.5/3.6).

As determined in Sect. 5.5.4 the total overshoot energy use of OCRT visitors is 127,310.4 GJ/year. Consequently the total cost to generate this amount of energy through using wind and solar systems can be calculated as follows:

127,310.4 GJ/year (overshoot portion of energy use) × NZ$ 72.90/GJ (the cost to generate 1 GJ energy through using wind + solar systems) = NZ$ 9,280,928.1

Table 5.5 shows that the total GDP of the OCRT is NZ$ 6,245,289. Consequently in the second scenario the remaining portion of GDP after subtraction of the cost to generate the overshoot portion of OCRT energy use (NZ$ 9,280,928.1) is NZ$ 6,245,289 – NZ$ 9,280,928.1= (–) NZ$ 3,035,639.1. This means that in the second scenario as well as total GDP an additional NZ$ 3,035,639.1 must be spent to restore the environmental impacts of the OCRT. What this study shows is, based on

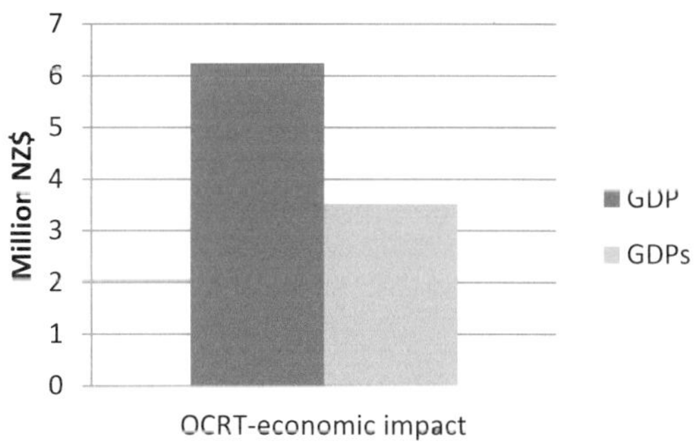

Fig. 5.1 Comparison between GDP and GDPs of OCRT

Table 5.8 New Zealand energy costs (second scenario)

Energy technology	Cost (NZ$/MWh)	Average cost (NZ$/MWh)
Wind	80–120[a]	100
Solar system	350–500[b]	425
Average cost of energy generated by wind-solar systems (NZ$/MWh)		262.5

[a]Reference: New Zealand Wind Energy association, 2011. (http://windenergy.org.nz/wind-energy/costs)

[b]Reference: www.Energywise.govt.nz

NZ$1 = US$0.83 (http://themoneyconverter.com), Viewed on 2011

the second scenario, the OCRT environmental loss is equivalent to NZ$ 3,035,639.1, something not revealed by its GDP. Comparison between the results of the two scenarios shows that using solar systems as a source of energy is environmentally friendly, but because of the high cost of this technology it is not currently a viable option when it comes to use of the GDPs comprehensive indicator. What the results also reveal is the OCRT, given the current price of renewable energy systems, cannot be considered as sustainable tourism when evaluated by a comprehensive indicator, like GDPs.

6 Conclusion

This study views economic development as a social-ecological phenomenon which is part of meeting the principles of sustainability. To overcome the shortcomings of current economic indicators with regard to sustainability, this article proposes using the sustainable portion of GDP (GDPs) as an indicator that has the ability to make a linkage between the ecological, social and economic dimensions of an activity, such as tourism and its related products and activities, as being sustainable. Through using the OCRT as a case study and using GDPs to explore the economic footprint of this tourism project, although the OCRT has been claimed by the New Zealand government as a successful sustainable tourism project, in a best cost assumption for using renewable resources for environmental remediation (first scenario) 43.6 % of total OCRT GDP must be spent to restore the environment damaged through development of the OCRT. Moreover, again using GDPs, but combining it with current renewable energy costs, this study shows that currently the OCRT is not sustainable.

References

Agenda21 (1992) United Nations Conference on Environment and Development (UNCED), Rio de Janeiro

Appleby P (2011) Integrated sustainable design of buildings. Lodon: Earthscan

Bagstad KJ, Shammin MR (2012) Can the genuine progress indicator better inform sustainable regional progress? – a case study for Northeast Ohio. J Ecol Econ 18:330–341

Barrass R, Sprenger RU, Triebswetter U (1997) Cohesion and the environment: policy integration at European level. J Eur Env 7:101–109

Burns G, Corbet R (2007) New Zealand Tourism strategy 2015. New Zealand Ministry of Tourism. Available at http://www.nztourismstrategy.com. Accessed on 13 Nov 2010

Central Otago District Council (CODC) (2011) OCRT user survey gross results. CODC, Otago

Daly HE, Cobb JB (1989) For the common good. Beacon Press, Boston

de Graaf HJ, Musters CJM, Keurs WJT (1996) Sustainable development: looking for new strategies. J Ecol Econ 16:205–216

Energy Efficiency and Conservation Authority (EECA) (2012) Available at. http://www.energywise.govt.nz, Accessed on 18 May 2012

Golusin M, Ivanovic OM (2009) Definition, characteristics and state of the indicators of sustainable development in countries of South-eastern Europe. J Agric Ecosyst Environ 130:67–74
Graham OJ (2009) The Otago Central Rail Trail: Preservation of heritage sites through development for visitor use, 3rd Australasian engineering heritage conference. Available at. http://www.ipenz.org.nz. Accessed on 12 Dec 2010
Jellum C, Reis A (2008) Otago Central Rail Trail economic impact survey, Otago Central Rail Trail Trust. Available at http://www.otagocentralrailtrail.co.nz. Accessed on 17 Dec 2010
Lawn PA (2003) A theoretical foundation to support the Index of sustainable economic welfare (ISEW), genuine progress indicator (GPI), and other related indexes. J Ecol Econ 44:105–118
Mahravan A (2012) Eco-tourism and its architecture: a methodological framework for assessing progress towards sustainable development. Published PhD thesis, Victoria University of Wellington
Mahravan A, Vale B (2011) Cultural framework and related indicators for evaluation of sustainability through development of tourism. Presented and published in Shanghai International Conference on Social Science (SICSS 2011), Shanghai
Mog JM (2004) Struggling with sustainability – a comparative framework for evaluating sustainable development programs. J World Dev 12:2139–2160
Nader MR, AbiSalloum B, Karam N (2008) Environment and sustainable development indicators in Lebanon: a practical municipal level approach. J Ecol Indicators 8:771–777
New Zealand Council of Trade Unions (NZCTU) (2010) Available at http://union.org.nz. Accessed on 17 May 2011
Nourry M (2008) Measuring sustainable development: some empirical evidence for France from eight alternative indicators. J Ecol Econ 67:441–456
Omer AM (2008) Energy, environment and sustainable development. J Renew Sust Energy Rev 12:2265–2300
Otago Central Rail Trial official web site, available at http://www.otagocentralrailtrail.co.nz/. Accessed on 21 May 2012
Pulselli FM, Ciampalini F, Tiezzi E, Zappia C (2006) The index of sustainable economic welfare (ISEW) for a local authority: a case study in Italy. J Ecol Econ 60:271–281
Rennings K, Wiggering H (1997) Steps towards indicators of sustainable development: Linking economic and ecological concepts. Ecol Econ 20:25–36
Robin CPY, Poon CS (2009) Cultural shift towards sustainability in the construction industry of Hong Kong. J Environ Manag 90:3616–3628
Spangenberg JH, Pfahi S, Deller K (2002) Towards indicators for institutional sustainability: Lessons from an analysis of Agenda 21. J Ecol Indicators 2:61–77
United Nations Environment Programme (UNEP) (2011) Towards a green economy, 100 Watt, France. Available at http://www.unep.org. Accessed on 28 Dec 2011
US Department of Energy (2011) (2010), Solar Technologies Market report D DOE/GO-102011-3318 NREL November, 60
Vale B, Vale R (2009) Time to eat the dog? Thames & Hudson Ltd., London
Vera I, Langlois L (2007) Energy indicators for sustainable development. Energy 32:875–882
Wackernagel M, Rees W (1996) Our ecological footprint: reducing human impact on the Earth. New Society Publishers, Gabriola Island
World Bank (2012) Poverty and equity data. Available at http://povertydata.worldbank.org/poverty/home. Accessed on 27 May 2012
World Bank Development Indicators (2008) Available at http://www.globalissues.org/issue/235/consumption-and-consumerism. Accessed on 16 Jan 2012
World Commission on Environment and Development (WCED) (1987) Brundtland report. Brundtland Commission

Chapter 6
Sustainable Transport: A Comparison of Ecological Footprint and Travel Patterns in Three Cities in Vietnam, New Zealand and Finland

Tran Thuc Han, Brenda Vale, and Robert Vale

Abstract There have been many debates on how to reduce the environmental impacts of travelling. Common suggestions are: increasing urban density, reducing travel distance to walkable and cyclable distances, using more public transport and so on. However, there is growing evidence which reveals that it is still unclear how urban form and density can be associated with travel distance. Hence, this article explores what forms the largest part of individual transport footprint in terms of domestic travel and the reasons why. Using ecological footprint measurement, this article will examine the environmental impact of domestic travel in three different cities: Hanoi in tropical Vietnam, Wellington in temperate New Zealand and Oulu in cold Finland. The argument is that the mode of travel is what should be the focus in order to reduce transport footprint, and high-density settlement is meaningless if travel still largely depends on the car. In the future of sustainability, people do not have to reduce travel distance within the city, but they do need to make smarter choices of travel.

Keywords Sustainable transport • Ecological footprint • Urban density • Travel mode • Travel distance

T.T. Han
Energy and Environmental Engineering Research Unit, Faculty of Technology, University of Oulu, Oulu, Finland

B. Vale • R. Vale (✉)
School of Architecture and Design, Victoria University of Wellington, Wellington, New Zealand
e-mail: robert.vale@vuw.ac.nz

T.H. Karyono et al. (eds.), *Sustainable Building and Built Environments to Mitigate Climate Change in the Tropics*, DOI 10.1007/978-3-319-49601-6_6

1 Introduction

Many cities have changed from the time the boundaries were defined by walking distance. This changed to cycling distance, and then extended to the tram era, and cities have sprawled from the time many people could access private motor vehicles. The environmental impact of transport has become very significant. There have been many debates on how to reduce the environmental impacts of travelling. Common suggestions are: increasing urban density, reducing travel distance to walkable and cyclable distances, using more public transport and electric vehicles and so on. But the question is how many people can afford to buy electric cars and how long would it take to replace the existing cars using fossil fuels? What are the alternative options in the transition time?

Urban sprawl is often assumed as unsustainable development that requires longer travel distances. The term "Compact city" suggests an opposite to urban sprawl that usually refers to higher density settlement through "attempts to increase built area and residential population densities" (Jenks and Burgess 2000, p. 9). A compelling study by Newman and Kenworthy (Rickwood et al. 2008), demonstrated that people living in higher density cities travel shorter distances, which means that they use less energy for transportation than those living in lower density cities. This idea of higher urban density reducing travel distance has been also supported by numerous scholars (Rickwood et al. 2008). However, there is growing evidence that it is still unclear how urban form and density can be associated with travel distance. Furthermore, there is the problem of how to ensure that people can find jobs within walkable and cyclable distances from their home. More importantly, having high density may also limit the opportunity for growing food and fuel for transportation, given the land area this needs (Vale and Vale 2010).

This article explores what forms the largest part of individual transport footprint (TF) in terms of domestic travel in three different cities, Hanoi in Vietnam, Wellington in New Zealand and Oulu in Finland, and the reasons why. Using ecological footprint (EF) measurement, this paper will examine the environmental impact of domestic travel in these three very different cities. This will be done through accessing EF of fuel uses, embodied energy (EE) of vehicles and infrastructure in the three cities.

In 2008, Hanoi was extended to become the Greater Hanoi region with over six million people and an area of 334,470 ha (Hanoi Urban Planning Institute 2009). The area of study for Hanoi in this research only covers the urban area – the old area of Hanoi that includes the nine main districts of Hanoi. Table 6.1 presents some significant differences between the three case study cities in terms of area, population, density and climate.

Table 6.1 Some comparisons of the three cities (Hanoi in Vietnam, Wellington in New Zealand and Oulu in Finland)

City	Hanoi (urban area only)	Wellington (Wellington City)	Oulu (area before January 2013)
Location	Capital of Vietnam	Capital of New Zealand	6th largest city in Finland
Established	1010	1840	1605
Area (ha)	17,900	290,000	151,300
Population	2,109,686 (year 2007)	201,898 (year 2013)	143,909 (year 2012)
Density (people/ha)	118	0.7	1.0
Annual average high temperature (°C)	27	16	6
Annual average low temperature (°C)	21	10	−2 (under 0)

Source:
Hanoi: area and population (Hanoi Urban Planning Institute 2009)
Wellington: area and population (WCC 2011)
Oulu: area and population (OUKA 2013)

2 Travel Patterns in the Three Cities

2.1 Mode of Travel

This section compares mode of travel in the three cities by looking at household vehicle ownership and share of trips by private vehicle, public transport and non-motorised transport.

There are significant differences in transport modes in the three cities in terms of private transport. The main means of private transport in Hanoi is the motorbike (commonly with a much smaller engine size than for motorbikes in the west, a motorbike in Vietnam would probably be categorised as a scooter in the other two countries). This is a much smaller type of vehicle than the car, which is significant in Wellington and Oulu. However, there is a similar low use of public transport (PT) systems in the three cities but for different reasons, as discussed below.

Table 6.2 describes household motor vehicle ownership in the three cities. It can be seen that the share of households with a motor vehicle in the three cities is very similar, at over 80% of the total. In fact, Hanoi has the highest share of motor vehicles per household of the three cities. Nevertheless, this changes when considering motor vehicles per person because the household size in Hanoi is about 4. In fact, having fewer people (less than 3) per household would limit the potential for full occupancy driving (in a car with 4–5 seats), which is necessary for fuel efficiency per person.

The share of PT in all travel in the three cities is low although there is a difference in the accessibility level of bus networks in these cities. The number of trip legs by bus in the three cities is similar, at 6% in Hanoi of all trips excluding walking

Table 6.2 Household motor vehicle ownership in Hanoi, Wellington and Oulu

Input elements	Hanoi	Wellington	Oulu
Year	2009	2006	2009
Main private motor vehicle type	Motorcycle	Car	Car
Household size (*)	3.9	2.55	2.1
Percentage of motor vehicles per household (%)(**)			
Total having motor vehicles	87.0	82.6	84.0
One motor vehicle	–	*44.7*	*53.0*
Two motor vehicles	*80% in new area* *50% in old area*	*29.3*	*29.0*
More than two motor vehicles	-	*8.6*	*2.0*
No motor vehicle	13.0	17.4	16.0

Source: Tran (2014, p. 159)

(Schipper et al. 2008, p. 8); 5% in Wellington (NZMT 2013); and only 4% in Oulu of all trips (FIMTC 2011). Surprisingly the share of trips made by bus in Wellington was very similar to Hanoi as well as Oulu although the public transport system in Wellington is very good and convenient. It is necessary to mention that using the bus in Oulu during the wintertime seems not an ideal option for many people due to the problem of waiting time under the sub-freezing temperatures. Furthermore, providing accessible bus services during the cold months is the issue, and would require more snow clearance during the long winter in order to provide a more accessible bus network in the city. Perhaps, together with providing accessible public transport, it is also necessary to change the travellers' preference choice (for example limiting car travel in the city centre and creating a more pedestrian friendly streetscape) in order to have greater use of public transport as well as more walking and cycling.

Currently, the use of the bus in Hanoi is mostly by low-income people, and it is not yet a popular means of public transport for office workers and students (Tran and Schlyter 2010). In the study survey, half of the respondents living in an old area that used the bus during the week to go to work were mainly factory workers with low incomes. Only 20% of those in the new area used the bus and most of the time the trips were for shopping. Furthermore, bus services in Hanoi seriously lack convenience due to the very high people loads in peak times and the fact that passengers are jostled during their trips.

The Wellington bus system is also unique because of its trolleybus network. After World War II, Wellington eventually chose trolleybuses to replace its trams, and now there are some 60 trolleys with Designline bodies and electrics from Brazil running in Wellington (IPENZ undated). The trolleybus system in Wellington is potentially a sustainable transport option in the future, once its electricity is generated from renewables.

Urban transport in Hanoi has changed dramatically since Vietnam opened up to the world economic network in the mid 1990s. Before that time, walking and cycling were the main means of travel within the city. Nowadays, the share of bicycle trips has decreased to about 20% in 2005, compared to 70% in 1995. Apart from very low-income groups, nowadays cycling is still the preferred transport mode for many students going to school and also for elderly people in Hanoi. Additionally, the share of walking trips in the city was 25% of the total trips (Schipper et al. 2008, p. 9).

Walking and cycling accounted for 16% of travel to work in Wellington city in 2001 (WCC 2005, p. 69). The share of these active modes in the wider Wellington region in 2006 was slightly smaller, at 13% of region-wide journey to work trips (WRC 2013b). Wellington seems a very walkable city as it is quite compact and concentrated. Wellington's CBD is about 2km in diameter and it takes about 20 min to walk from one side to the other (WCC 2010). Hence, this offers pedestrians a great chance to walk to work or to combine walking trips with its well-connected public transport. However, the city landscape is rather hilly for cycling. Another reason that could deter cyclists is that there are few separate cycling lanes, and most cyclists have to use the same road as cars. For these reason, currently cycling does not seem a very safe and user-friendly travel option in Wellington. As a result, cycling accounted for only 1% of all trip legs whereas pedestrians accounted for 25% (NZMT 2013).

In contrast, although the wintertime is tough in Oulu, the share of walking and cycling is relatively high there, at 15% of the trips for each mode, 30% in total (FIMTC 2011). There are several reasons for Oulu's cycling success. First, unlike Wellington, Oulu is a flat city. Second, it provides a very good network for cycling with separate lanes, as well as good winter maintenance. Last but most important, cycling and also winter cycling are part of its historic cultural activities and the citizens of Oulu are very proud of their reputation as winter cyclists. However, there has still been a significant reduction in the share of non-motorized transport modes. In 1962, the shares of walking and cycling in the total transport mode choice were 27% and 42% respectively, making the non-motorized mode share 69% of the total. In 1989, these had reduced to 16% for walking and 28% for cycling. Now the share of walking has slightly increased to 19%, but the share of cycling has continued to decline to 21% (Kati et al. 2010). Perhaps providing a good network for bicycles is not enough in encouraging people to cycle, unless there is a limitation in car driving at the same time.

2.2 *Travel Distance*

As can be seen from Table 6.3, the travel distances per person per year in the three different cities are actually very similar, ranging from 11,008 to 12,187 km/person/year. The difference between the cities is seen in the ways that people choose to travel presented in terms of transport mode. In particular, Wellington has the highest distance travelled by public transport at 823 km/person (accounting for 7.1% of

Table 6.3 Annual distance travelled per person (km/person) by mode excluding long distance travel by train and air travel, in Hanoi, Wellington and Oulu

Year	2005	2008–2011	2009
Distance by mode	Hanoi (km/person/year)	Wellington (km/person/year)	Oulu (km/person/year)
Public transport	**219**	**823**	**290**
Bus	219	290	290
Trolley bus	–	*112*	–
Diesel bus	–	*178*	–
Train (local only)	–	533	–
Ferry	–	–	–
Private transport	**7397**	**10,280**	**10,950**
Car driver	789	6486	9239
Car passenger		3794	1711
Motorbike driver	5905	–	–
Motorbike passenger	703		
Non-motorised	**3282**	**390**	**802**
Cycling	2237	60	620
Walking	1045	330	182
Other	**110**	**115**	**145**
Total	**11,008**	**11,608**	**12,187**

Source: Tran (2014, p. 170)

total distance) because besides the bus network, the city also has its suburban railway network.

The travel distances by car in Wellington and Oulu are very similar, over 10,000 km/per/year (amounting to almost 90% of total distance), but there was more car sharing in Wellington compared to Oulu, and the distance travelled by car passengers in Wellington was more than double that of Oulu (3794 compared to 1711 km/person/year). Within the three cities, Hanoi has the lowest distance travelled by motorised private vehicle, 7397 km/person/year (789 km car, 5905 km as motorbike driver and 703 km as motorbike passenger). The least sharing is seen in private vehicle use in Hanoi (6.4% total distance travelled), perhaps not surprising given the fact that a motorbike is already a small vehicle that suits 1 or 2 persons travelling, although in other Asian countries, including China, Indonesia and Thailand, four people on a small motorbike – 2 adults and 2 children – is a common sight.

Hanoi also has the highest distance share of non-motorised travel with over 3000 km/person/year (over 2000 km cycling and 1000 km walking) that accounted for almost 30% of total travel distance (Table 6.4). Cycling is still the main means of transport for students and older groups. Female workers cycled more than men, 28% compared to 7% (Nguyen 2011, p. 130). Perhaps cycling would appear to be a more suitable mode choice for women in Hanoi as women travelled shorter distances than men (Tran and Schlyter 2010) and usually cycling is safer than driving a motorcycle. Nevertheless, apart from the safety factor, cycling is a more affordable second vehicle option for many families having only one motorbike in Hanoi.

Table 6.4 Share of distance travelled (%) by mode of transport in Hanoi, Wellington and Oulu

Year	2005	2008–2011	2009
Mode share of distance	Hanoi (distance %)	Wellington (distance %)	Oulu (distance %)
Public transport	**2.0**	**7.1**	**2.4**
Bus	2.0	2.5	2.4
Train	–	4.6	–
Private transport	**67.2**	**88.6**	**89.8**
Car driver	7.2	55.9	75.8
Car passenger		32.7	14.0
Motorbike driver	53.6	–	–
Motorbike passenger	6.4		
Non-motorised	**29.8**	**3.3**	**6.6**
Walk	9.5	2.8	1.5
Cycle	20.3	0.5	5.1
Others	**1.0**	**1.0**	**1.2**
Total	**100**	**100**	**100**

Source: Tran (2014, p. 171)

Compared to Hanoi, the distance shares of non-motorised travel in Wellington and Oulu are much lower at 3.3% and 6.6% respectively. Although Oulu has a very good network for cycling and is considered to have the most active bicycling culture in Finland (Business Oulu 2011), cycling distance is still much lower than in Hanoi, 620 km compared to 2237 km/person/year. Perhaps the reason is cycling is more limited during the long dark winter in Oulu and the fact that access to a motorised private vehicle (car) per person in Oulu is more available than in Hanoi (motorbike) given the Oulu household size is approximately 2 and that of Hanoi is about 4 (Table 6.2). Many people cycle in Oulu because they enjoy it and cycling is also good exercise, whereas many people in Hanoi have to cycle because they do not have other choices due to their low income and poor public transport system.

Walking amounted to almost 10% of total distance in Hanoi compared to less than 3% in Wellington and 1.5% in Oulu (Table 6.3). The reason is perhaps that Hanoi is a compact city in terms of high intensity of mixed-use that provides more opportunities for local shopping and services (Tran 2008). In other words, shops and services in Hanoi are very available within walkable and cyclable distances and such patterns of settlement provide more opportunities for people to choose non-motorised travel.

2.3 *Travel by Purpose*

Tables 6.5 and 6.6 compare travel distance by purpose in km/person/year and proportion of travelled kilometres (%) for the three case studies. The results reveal that commuting to work and travel associated with business accounted for the longest distance of travel in all cities. It was 30% (3533 km) and 34% (4144) of the total

Table 6.5 Comparison of travel distance by purpose (km/person/year), in Hanoi, Wellington and Oulu

Purpose	Hanoi (km)	Wellington (km)	Oulu (km)
Work and business	4954	3533	4144
Work-main job	–	*1921*	*2925*
Work-employers business	–	*1612*	*1219*
Education	826	276	853
Shop and services	1706	3019	2193
Shopping	–	*1818*	*1462*
Personal business	–	*1201*	*731*
Social-recreational	2532	3155	3169
Visiting	–	–	*1341*
Recreational	–	–	*1828*
Accompany someone else	330	1098	609
Other (*)	660	527	1219
Total	**11,008**	**11,608**	**12,187**

Source: Tran (2014, p. 180)

Table 6.6 Comparison of share of distance travelled by purpose (%), in Hanoi, Wellington and Oulu

Purpose	Hanoi (%)	Wellington (%)	Oulu (%)
Work and business	45	30.4	34
Work-main job	–	*16.5*	*24*
Work-employers business	–	*13.9*	*10*
Education	7.5	2.4	7
Shop and services	15.5	25.9	18
Shopping	–	*15.6*	*12*
Personal business	–	*10.3*	*6*
Social-recreational	23	27.2	26
Visiting	–	–	*11*
Recreational	–	–	*15*
Accompany someone else	3	9.5	5
Other (*)	6	4.6	10
Total	**100**	**100**	**100**

Source: Tran (2014, p. 180)

travel distance for Wellington and Oulu respectively and accounted for almost half the travel distance for someone living in Hanoi (45%). The distance travelled for work/business in Hanoi was the longest of the three case studies, almost 5000 km/person/year. This is possibly because due to low incomes, it is very common that people in Hanoi have more than one source of employment, and hence people have to travel more for their businesses. Mixed-use housing shared between living and shop/service in Hanoi is one of the examples showing that people are having more than one business (Tran 2013).

In contrast, within the three case studies, distances travelled for shopping and services per person in Hanoi are the shortest at 1706 km compared to Wellington 3019 km and Oulu 2193 km (see Table 6.5). This does not mean that people in Hanoi do less shopping than in Wellington and Oulu. In fact, women in Hanoi go to the shops for food more frequently and many families in Hanoi still prefer buying fresh food every day. The main reason here is because shopping trips can largely be done locally. Again, referring to the case of mixed-use housing in Hanoi that provides local shops and services, most daily shopping is within walkable and cyclable distances. This is very different from how shopping is done in Wellington and Oulu, where usually shopping for necessities such as groceries is done by car, one or two times per week with longer distances to a big supermarket, and sometimes to more than one when they have special deals.

Social-recreational trips are another reason for longer distance travel although these may not be as frequent as work, shopping or services trips. Most of the time people travel to visit friends/relatives or to get out of the cities, normally during the weekend. There was not much difference within the three cities in terms of social-recreational travel distances, as these ranged from 2500 km/person/year-about 23% of total travel distance (Hanoi) to over 3000 km/person/year-about 25% of total travel distance (Wellington and Oulu). In short, the need for social life and leisure activities appears quite similar in the three cities. However, in this context, data for social-recreational travel do not cover the long-distance travel often involved in social/recreational trips. There are no available data for long distance social-recreational travel in Hanoi and Wellington. However, there are available data for all social-recreational travel in the Finland National Travel Survey (Liikenennevirasto and WSP Finland 2012). It states that in 2010–2011, the average travel distance per person per day for visiting trips was almost 8 km, for leisure trips was 10 km, and for summer cottage trips was 2.2 km. This means, the average travel distances for visiting and leisure per person per year in Finland were 2920 km and 3650 km and for going to the summer cottage was 800 km; making a total of 7370 km/person/year for domestic social/recreational trips. This means social/recreational trips were the largest part of travel in Finland.

Although leisure travel is infrequent travel it has been a major component of travel undertaken in New Zealand over the last forty years (Dravitzki et al. 2009). More importantly, Dravitski et al analysed the link between social/recreational trips and the uptake of private vehicles. The private vehicle allows people to travel to a greater range of destinations than public transport, which highlights that much social/recreational travel takes people outside the urban area, meaning that one's local community is now separated in space and linked by transport (especially cars) and telecommunication.

Although trips made for an education purpose are frequent and often every weekday their share of total distance is relatively low, being 800 km/person/year (about 16–18% of total travel distance) in both Hanoi and Oulu, and just less than 300 km/person/year in Wellington.

Table 6.7 shows the level of dependence on the private vehicle in the three cities for different destinations. Data for Wellington are the adjusted national data of New

Table 6.7 Comparison of share of trips (%) made by private motorised vehicle, motorbike in Hanoi and car in Wellington and Oulu, by travel purpose

Year	2010	1997–1998	2009
Purpose	Hanoi (%)	Wellington (%)	Oulu (%)
Type of private vehicle	**Motorbike**	**Car**	**Car**
Work and business	**60**	**82**	**75**
Work-main job	*60*	*76*	*68*
Work-employers business	–	*88*	*81*
Education	**28**	**47**	**18**
Shop and services	**51**	**77**	**67**
Shopping	*60*	*76*	*64*
Groceries	*29*	–	–
Personal business	*65*	*78*	*70*
Social-recreational	**45**	**73**	**62**
Visiting	*51*	–	*60*
Recreational	*39*	–	*63*
Accompany someone else	**89**	**91**	**92**

Source: Tran (2014, p. 181)

Zealand, as there are no available data for the city in this case. It shows that Wellington has the highest share of trips made by car for all purposes although distance for car travel per person per year in Wellington was about the same as in Oulu (see Table 6.3). Perhaps the share of trips made by car in Wellington would be slightly lower if local data were available; as mentioned above, the share of trips made by public transport in Wellington was slightly higher than the national average (5% compared to 3%). Nevertheless, there are significant differences in choices people make to travel to different destinations in the three cities, though the distance travelled per person per year in the three cities is similar. Hanoi has the lowest share of trips made by private motorised transport (motorbike) for almost all travel purposes excluding education trips.

In fact, the share of trips made by private motorised transport (car) for education is lowest in Oulu at 18% of total trips by purpose. Although the average distance travelled to school in Wellington is the lowest within the three cities, at less than 300 km/person/year (see Table 6.5), 47% of education trips were by private motorised transport (car). Not surprisingly, the share of private motorised vehicles for trips accompanying someone else is about 90% in all three cities.

3 Higher Density?

As long as households have a car, they seem to drive further for shopping and for leisure trips during the weekends and holidays. Travel to the summer cottage in Finland is an example, as this has become more and more common, especially with urban dwellers, since many families have cars and they can access their summer

Table 6.8 Travel distance (km/person/day) by purpose according to household car ownership, Finland, year 2004–2005

Purpose	No cars	1 car	2 cars	Over 2 cars	Average
Work	2.9	6.8	10.5	11.5	7.4
Business	0.7	3.4	6.0	8.3	3.9
Education	0.9	1.2	2.3	4.9	1.6
Shopping and services	2.8	5.9	6.7	8.5	5.8
Visits	9.0	8.8	8.6	8.2	8.7
Summer cottage	1.0	3.0	2.8	3.8	2.6
Other Leisure	5.5	12.0	14.5	15.5	11.8
Total	**22.8**	**41.2**	**51.4**	**60.7**	**41.8**

Source: Liikenennevirast and WSP Finland (2006) and Tran (2014, p. 183)

cottages more often. Many Finns are apartment dwellers, considered as high-density settlement, but adapting socially and emotionally to this urban landscape has been problematic for many migrants from the rural areas. Many urban dwellers still view themselves as partly rural and there is a need to return to summer cottages in the rural areas during leisure times (cited in everyculture 2006). There has been much discussion about urban form and density to reduce travelling, such as the Wellington Transport Strategy that is aimed to find a stronger linkage between urban form and the transport needed to support it (WCC 2005, p. 72). However, it is still unclear how urban form and density can be associated with travel distance.

This research into travel patterns of three cities reveals that people still travel similar total annual distances regardless of huge differences in city density. What differs is their choice of travel mode. However, there is evidence that the car influences the way that people choose to travel and the distance travelled. Table 6.8 shows clearly that, in Finland, the more cars people have, the further they travel: people having no cars travel only about 22 km/day whereas those from households having 1 car travelled double the distance (41.2 km), with 2 cars 51.4 km and more than 2 cars 60.7 km, and most of these distances were by car. Hence, the key questions are how to reduce private vehicle travel and to what extent people should have access to a car or motorbike, neither of which appears to be clearly linked to density.

It seems it would be difficult to reduce the distance travelled to work, business, and for education and visits, although it could be possible to change the travel mode for these purposes. However, it is possible to reduce the distance travelled for shopping, services and leisure activities, which form a significant part of the total, as well as the way people get there. In Hanoi, before the economic boom at the end of the 1990s, leisure trips were trips every weekend to visit parents and long-distance travel was not common. Long-distance travel only happened once a year during the school summer holiday and most trips were taken by train to go to the beaches close to Hanoi. Also there were not many shopping malls to go to during the weekend. Apart from daily shopping for necessities such as food, most people usually shopped for new goods only in the New Year period or special events. Even now many people in Hanoi do not undertake long distance travel as frequently as in Wellington or Finland. However, this way of living has been changed among the younger generation as the consequence of globalisation.

4 Transport Footprints

This section estimates the TF or transportation footprint in terms of vehicle embodied energy (EE), infrastructure EE and fuel use according to mode of vehicle. Finally, based on the total TF result, the research provides the EF of transportation according to travel purpose. In estimating the footprint of the current transport system that is largely based on fossil fuel, the study uses the general energy yield factor of 100 GJ/ha/year (Wackernagel and Monfreda 2004, p. 6).

Table 6.9 is produced to estimate the EF of vehicle EE per person in the three cities. Not surprisingly, the result reveals that vehicle EE per person of Hanoi (0.0263 ha/person) is lowest, and the vehicle footprints of Wellington and Oulu are similar (0.0546 and 0.0615 ha/person). The reason is people in Hanoi use smaller and more compact private vehicles (motorbike) and more trips are made by cycling and walking than in Wellington and Oulu.

Table 6.10, the footprint of infrastructure EE (ha/person), shows the infrastructure footprints are very similar in all three cities (over 0.04 ha/person) regardless of the choice of mode of travel. It means nothing much can be done to reduce the environmental impact of infrastructure in terms of passenger mode choice. Perhaps, people could use more PT to replace private vehicles as the infrastructure EE of the bus is only 0.2 MJ/km.

Table 6.11 gives the EF relating to fuel use for passenger transport per person in the three cities, in terms of land area needed to produce the fuel. The fuel footprints per person in the three cities are very different. The footprint of someone living in Hanoi (0.1171 ha/person) is less than half of someone in Wellington (0.2869 ha/person), and about one third of someone in Oulu (0.3424 ha/person). This suggests that using more compact vehicles with more cycling and walking would significantly reduce the environmental impact of passenger transport. The reason why the TF of Wellington (travel as car passenger in Wellington is 3794 km compared to Oulu 1711 km) is smaller than Oulu is because there is more car sharing in Wellington although the distances travelled by car per person in the two cities are almost the same. Moreover, the use of PT in Wellington is higher than in Oulu due to its local train network.

Finally, using the footprint results in Tables 6.9, 6.10 and 6.11, Table 6.12 presents the total land area needed for personal transportation in Hanoi, Wellington and Oulu (ha/person/year) including EE for vehicles and infrastructure and fuel consumption. According to the results, the footprint of fuel per person per year is much higher than the total EE of vehicles and infrastructure, in Hanoi it is two times higher; Wellington and Oulu are both about three times higher. Although a similar infrastructure footprint is seen in all three cities, the overall TF (fuel, vehicles, infrastructure)

Table 6.9 Estimated EF of vehicle EE per person in Hanoi, Wellington and Oulu (ha/person/year)

Year	2005	2008–2011	2009		Energy used per person (GJ/person)			EF of vehicle per person (ha)		
Distance by mode (km/person/year)	Hanoi	WLG	Oulu	Vehicle EE (MJ/km)	Hanoi	WLG	Oulu	Hanoi	WLG	Oulu
PT	**219**	**823**	**290**		**0.055**	**0.179**	**0.072**	**0.0005**	**0.0018**	**0.0007**
Bus	219	290	290	0.25	0.055	0.072	0.072	0.0005	0.0007	0.0007
Train (local only)	–	533	–	0.20	–	0.107	–	–	0.0011	–
Private transport	**7397**	**10,280**	**10,950**		**2.112**	**5.140**	**5.475**	**0.0211**	**0.0514**	**0.0547**
Car	789	10,280	10,950	0.50	0.394	5.140	5.475	0.0039	0.0514	0.0547
Motorbike	6608	–	–	0.26	1.718	–	–	0.0172	–	–
Non-motorised	**3282**	**390**	**802**		**0.440**	**0.087**	**0.095**	**0.0044**	**0.0008**	**0.0054**
Cycling	2237	60	620	0.08	0.179	0.005	0.050	0.0018	0.00005	0.0050
Walking	1045	330	182	0.25	0.261	0.082	0.045	0.0026	0.0008	0.0004
Other	**110**	**115**	**145**	Car taxi 0.50 Motorbike taxi 0.26	**0.029**	**0.057**	**0.072**	**0.0003**	**0.0006**	**0.0007**
Total	**11,008**	**11,608**	**12,187**		**2.636**	**5.463**	**5.714**	**0.0263**	**0.0546**	**0.0615**

Source: Tran (2014, p. 189)
Note: WLG is Wellington

Table 6.10 Estimated EF of infrastructure EE per person in Hanoi, Wellington and Oulu (ha/person/year)

Year	2005	2008–2011	2009		Energy used per person (GJ/person)			EF of infrastructure per person (ha)		
Distance by mode (km/per/year)	Hanoi	WLG	Oulu	Infra EE (MJ/km)	Hanoi	WLG	Oulu	Hanoi	WLG	Oulu
PT	**219**	**823**	**290**		**0.044**	**0.271**	**0.058**	**0.0004**	**0.0027**	**0.0006**
Bus	219	290	290	0.2	0.044	0.058	0.058	0.0004	0.0006	0.0006
Train (local only)	–	533	–	0.4	–	0.213	–	–	0.0021	–
Motorised private transport	**7397**	**10,280**	**10,950**	0.4	**2.959**	**4.112**	**4.380**	**0.0296**	**0.0411**	**0.0438**
Non-motorised	**3282**	**390**	**802**	0.4	**1.313**	**0.156**	**0.321**	**0.0131**	**0.0016**	**0.0032**
Other	**110**	**115**	**145**	0.4	**0.044**	**0.046**	**0.058**	**0.0004**	**0.0005**	**0.0006**
Total	**11,008**	**11,608**	**12,187**		**4.360**	**4.585**	**4.817**	**0.0435**	**0.0459**	**0.0482**

Source: Tran (2014, p. 189)

Note: It is assumed in Table 6.10 that each mode has its own infrastructure EE, but in practice 'motorized private transport', 'non-motorized' and 'other' to some extent share their infrastructure. WLG is Wellington. "Infra" is infrastructure

Table 6.11 Estimated EF of fuel use (FU) in passenger transport per person in Hanoi, Wellington and Oulu (ha/person/year)

Year	2005	2008–2011	2009	Energy used per km (MJ/km)	Energy used (GJ/person)			EF of FU per person (ha)		
Distance by mode (km/per/year)	Hanoi	WLG	Oulu		Hanoi	WLG	Oulu	Hanoi	WLG	Oulu
PT	**219**	**823**	**290**		**0.77**	**1.22**	**1.01**	**0.0077**	**0.0122**	**0.0101**
Bus	219	290	290	3.50	0.77	0.91	1.01	0.0077	0.0091	0.0101
Trolley bus	–	*112*	–	*2.61*	–	*0.29*	–	–	*0.0029*	–
Diesel bus	–	*178*	–	*3.50*	–	*0.62*	–	–	*0.0062*	–
Train (local only)	–	533	–	0.59 (electric)	–	0.31	–	–	0.0031	–
Ferry	–	–	–	–	–	–	–	–	–	–
Private transport	**7397**	**10,280**	**10,950**		**8.77**	**26.76**	**32.21**	**0.0877**	**0.2676**	**0.3221**
Car driver	789	6486	9239	3.19	2.52	20.69	29.47	0.0252	0.2069	0.2947
Car passenger		3794	1711	1.60		6.07	2.74		0.0607	0.0274
Motorbike driver	5905	–	–	1.00	5.90	–	–	0.0590	–	–
Motorbike passenger	703			0.50	0.35			0.0035		
Non-motorised	**3282**	**390**	**802**		**2.06**	**0.34**	**0.56**	**0.0206**	**0.0034**	**0.0056**
Cycling	2237 (slow)	60 (fast)	620 (fast)	(fast) 0.65 (slow) 0.50	1.12	0.04	0.40	0.0112	0.0004	0.0040
Walking	1045	330	182	0.90	0.94	0.30	0.16	0.0094	0.0030	0.0016
Other	**110**	**115**	**145**		**0.11**	**0.37**	**0.46**	**0.0011**	**0.0037**	**0.0046**
Moped taxi				1.00	0.11			0.0011	–	–
Car taxi				3.19		0.37	0.46	–	0.0037	0.0046
Comparison: average mix				*1.90*	*0.21*	*0.22*	*0.27*	*0.0021*	*0.0022*	*0.0027*
Total	**11,008**	**11,608**	**12,187**		**11.60**	**28.69**	**34.24**	**0.1171**	**0.2869**	**0.3424**

Source: Tran (2014, p. 191)

Table 6.12 Estimated total EF of transportation excluding long distance travel by aircraft and train, per person (ha/person/year) in Hanoi, Wellington and Oulu

Year	2005	2008–2011	2009
City	**Hanoi**	**Wellington**	**Oulu**
Area (ha)	17,900	247,700	151,300
Population (people)	2,109,686	201,891	143,909
Household-size (people)	3.90	2.55	2.1
EF of FU (ha/year)	0.1171	0.2869	0.3424
EF of vehicle (ha/year)	0.0263	0.0546	0.0615
EF of infrastructure (ha/year)	0.0435	0.0459	0.0482
Total EF of transportation per person (ha/person/year)	**0.1869**	**0.3874**	**0.4521**
Total EF of transportation per household (ha/household/year)	**0.7289**	**0.9879**	**0.9494**
Total transport EF of the whole city (ha/year)	**394,300**	**78,212**	**65,061**

Source: Tran (2014, p. 192)

of Hanoi (0.19 ha/person) is much lower than Wellington (0.39 ha/person) and Oulu (0.45 ha/person). The case studies of the three cities show it is the mode of vehicle that should be the focus in order to reduce the land area footprint needed for individual transportation.

At the level of the household TF, interestingly, this is almost the same in all three cities (the range is 0.7–1.0 ha/household) due to the number of persons per household. Nevertheless, the household TF of Hanoi is still the lowest despite the larger household size. However, at the city scale, the total land area needed for transportation for all citizens in Hanoi is more than 22 times the city's land area. Again, this brings up the question whether having a high-density settlement is sustainable if sustainability means living within what can be produced sustainably within the available land area. The land area needed for transport in Wellington is about one third of the area of the city, and for Oulu about half the city size.

Table 6.13 presents EF of personal transport by purpose to see what the largest part of travel is in the three case studies. The result reveals that work, shopping and recreation form most travel in all three cities. The footprints for work and business travel are similar in the three cities (about 0.1 ha/person), and the major difference is seen in the way people go to shop and enjoy their leisure time.

The case of Hanoi shows that the TF for work and business accounted for the largest part of the footprint at almost half the total, due to the fact people have more than one business, and the second largest part is social recreation at 23%, whereas the footprint for shopping is relatively low (15.5%) because it is largely local shopping. Similarity is seen in the two cities of Wellington and Oulu, where the needs for work, shopping and recreation are almost the same (around 0.1 ha/person). The main reason for Wellington and Oulu having a much higher shopping and services footprint (over 3 times higher than Hanoi) is because shopping trips are largely by car to big supermarkets instead of local shops. The Wellington and Oulu social-

Table 6.13 Estimated EF of transportation by purpose per person (ha/person/year) in Hanoi, Wellington and Oulu

	Share of distance travel (%)			EF by travel purpose (ha/person/year)		
Purpose	Hanoi	Wellington	Oulu	Hanoi	Wellington	Oulu
Work and business	45	30.4	34	**0.0841**	**0.1178**	**0.1537**
Work-main job	–	*16.5*	*24*	–	*0.0639*	*0.1085*
Work-employers business	–	*13.9*	*10*	–	*0.0539*	*0.0452*
Education	7.5	2.4	7	**0.0140**	**0.0093**	**0.0317**
Shop and services	15.5	25.9	18	**0.0290**	**0.1003**	**0.0814**
Shopping	–	*15.6*	*12*	–	*0.0604*	*0.0543*
Personal business	–	*10.3*	*6*	–	*0.0399*	*0.0271*
Social-recreational	23	27.2	26	**0.0430**	**0.1054**	**0.1175**
Visiting	–	–	*11*	–	–	*0.0497*
Recreational	–	–	*15*	–	–	*0.0678*
Accompany someone else	3	9.5	5	**0.0056**	**0.0368**	**0.0226**
Other (*)	6	4.6	10	**0.0112**	**0.0178**	**0.0452**
Total distance travel (km/person/year)	**11,008**	**11,608**	**12,187**	**0.1869**	**0.3874**	**0.4521**
Ratio	**1.00 reference point**	**1.05**	**1.10**	**1.00 reference point**	**2.07**	**2.4**

Source: Tran (2014, p. 193)
(*): Other trips refer to medical, change mode, day care, etc.

recreational footprints are both double that of Hanoi because many leisure trips in Hanoi are by smaller vehicle (motorbike) within the city to visit relatives and friends, whereas people in Wellington and Oulu usually drive to get out of the city or drive to shopping malls.

The results in Table 6.14 suggest although TF per capita in Hanoi is much smaller than for a resident of Wellington or Oulu, it is impossible for Hanoi to generate energy for transport at the city level. On the other hand, it would be possible to do so in the other two cities based on their land area.

5 Conclusion

This investigation of transport footprint reveals that people in the three cities still travel almost the same distances regardless of huge differences in urban densities as well as urban forms. Instead, what differs is their choice of travel mode. This leads to significant differences in individual transport footprint results for the three cities. The individual transport footprint of Hanoi is much smaller than those of Wellington and Oulu, although the travel distances are very similar.

Table 6.14 Comparison total TF of the three cities (ha)

Year	2005	2008–2011	2009
City	**Hanoi**	**Wellington**	**Oulu**
Area (ha)	17,900	247,700	151,300
Population (people)	2,109,686	201,891	143,909
Household-size (people)	3.90	2.55	2.1
EF of FU (ha/year)	0.1171	0.2869	0.3424
EF of vehicle (ha/year)	0.0263	0.0546	0.0615
EF of infrastructure (ha/year)	0.0435	0.0459	0.0482
Total EF of transportation per person (ha/person/year)	**0.1869**	**0.3874**	**0.4521**
Total EF of transportation per household (ha/household/year)	**0.7289**	**0.9879**	**0.9494**
Total TF of the whole city (ha/year)	**394,300**	**78,212**	**65,061**

Source: Tran (2014, p. 194)

This result of the study suggests that the mode of travel is what should be the focus in order to reduce transport footprint, and high density is meaningless if travel still largely depends on the car. It also argues that in the future of sustainability, people do not have to reduce travel distance within the city, but need to make smarter choices of travel. It means people do not necessarily have to work at home, buy stuff online, or contact or see each other via Internet networks. Instead, they can switch to more compact private vehicles such as motorbikes or use public transport. Rather than driving the car to big supermarkets, they should walk and cycle to local shops. And in terms of leisure trips, people should visit their friends and relatives during the weekend, rather than driving to shopping malls. Indeed, environmental impact of travelling can be reduced significantly through making smarter choices of travel. To do so requires change in behaviour but most importantly, it can be done within the existing physical environment.

References

Business Oulu (2011) High quality of living [Online]. Available: http://www.businessoulu.com/en/oulu/living-environment-with-quality-and-opportunities.html. Accessed Mar 2013

Dravitzki V, Lester T, Walton D (2009) Social/recreational travel and its influence on transport's greenhouse gas emissions [Online]. Australasian Transport Research Forum. Available: http://www.atrf.info/papers/2009/2009_Dravitzki_Lester_Walton.pdf. Accessed May 2013

EVERYCULTURE (2006) Culture of Finland [Online]. Available: http://www.everyculture.com/Cr-Ga/Finland.html-b. Accessed Dec 2012

FIMTC (2011). Oulu region travel survey 2009 [Online]. Oulun kaupunki - City of Oulu Office. Available: http://oulu.ouka.fi/seutu/pdf/OulunSeudunLT2009esittely_eng.pdf. Accessed Feb 2013

Hanoi Urban Planning Institute (2009) Thuyết Minh Quy Hoạch Chung Xây Dựng Thủ đô Hà Nội đến năm 2030 và Tầm Nhìn đến năm 2050 (Master Plan for the Capital Hanoi to 2030 and Vision to 2050)

IPENZ (undated) Wellington trolley buses, 1924–1932 and 1949–to date [Online]. Engineering Heritage New Zealand. Available: http://www.ipenz.org.nz/heritage/itemdetail.cfm?itemid=2257. Accessed Mar 2013

Jenks M, Burgess R (2000) *Compact cities: sustainable urban forms for developing countries*. Spon Press, London

Kati K, Ylinampa J, Heikkinen J, Hivornen M (2010) The secret of winter cycling's success in Oulu. Velo-city Global 2010. Copenhagen

Liikenennevirast, WSP Finland (2006) The National Travel Survey 2004–2005 [Online]. Finnish Transport Agency and WSP LT Consultants Ltd. Available: http://www2.liikennevirasto.fi/hlt20042005/english/modes.htm. Accessed Mar 2013

Liikenennevirasto, WSP Finland (2012) National Travel Survey 2010–2011. Finnish Transport Agency and WSP LT Consultants Ltd

Nguyen HTT (2011) Élements Pour Une Mobilite Quotidienne Compatible Avec Le Transport Durable Au Vietnam: Enjeux Et Perspectives D'un Report Modal Vers Les Transports Collectifs Et Les Transports Non Motorises, Le Cas De Hanoi (Factors for a Daily Mobility Compatible with Transport Sustainable in Vietnam: Challenges and Prospects for a Modal Shift Towards Public Transport and Non-Motorized Transports, The Case of Hanoi). Doctoral degree, National Institute of Applied Science, Lyon

NZMT (2013) Latest results – New Zealand Househols Travel Survey: Regional Travel Results [Online]. Available: http://www.transport.govt.nz/research/LatestResult/. Accessed Feb 2013

OUKA (2013) Information about Oulu: population and area [Online]. Available: http://oulu.ouka.fi/english/general/informationaboutoulu/population.htm. Accessed Jan 2013

Rickwood P, Glazebrook G, Serle G (2008) Urban structure and energy: a review. *Urban Policy Research* 26(1):57–81

Schipper L, Anh TL, Orn H (2008) Measuring the invisible: quantifying emissions reductions from transport solutions [Online]. World Resources Institute. Available: http://pdf.wri.org/measuringtheinvisible_hanoi-508c_eng.pdf. Accessed Dec 2011

Tran HA, Schlyter A (2010) Gender and class in urban transport: the case of Xian and Hanoi. *Environment and Urbanization* 22:139–155

Tran HT (2008) Asian cities and compact urban form: a study of Hanoi. Master of Sustainable Design, The University of Auckland

Tran HT (2013) A study of Hanoi. In: Vale R, Vale B (eds) *Living within a fair share ecological footprint*. Routledge, London

Tran HT (2014) Sustainable patterns of living based on an investigation of footprint in Hanoi-Vietnam, Wellington-New Zealand and Oulu-Finland. Doctor of Philosophy in Architecture, Victoria University of Wellington

Vale B, Vale R (2010) Is the high-density city the only options? In: Ng E (ed) *Designing high-density cities for socila and environmental sustainability*. Earthscan, London

Wackernagel M, Monfreda C (2004) Ecological footprints and energy. In: *Encyclopedia of energy*, vol 2. Elsevier Academic Press, Amsterdam

WCC (2005) Wellington City Profile and Policy Stocktake

WCC (2010) Compact and connected [Online]. Available: http://live.wellingtonnz.com/page/compact-and-connected.aspx. Accessed Mar 2013

WCC (2011) About Wellington: facts and figures [Online]. Available: http://www.wellington.govt.nz/aboutwgtn/glance/index.html. Accessed Jan 2013

WRC (2013) Walking [Online]. Available: http://www.gw.govt.nz/walking-4/. Accessed Mar 2013

Chapter 7
Smart Community Energy Systems for Low Carbon Living

A.D. Peacock, E.H. Owens, Sue Roaf, and D.W. Corne

Abstract In a world where the need to Mitigate against, and Adapt to, a rapidly changing climate is pressing, so too is the need to re-think the way we power our lifestyles. If we are to survive as a species we will need to substantially reduce the greenhouse gas emissions that result from our life-styles. In addition we will need to develop energy supply systems that can cope with more extreme weather conditions, keep us safe through power outages on the traditional grid systems and provide us with affordable comfort in our everyday lives. When building visions for the best possible future energy systems in a world with growing populations, limited fossil fuel resources, rising energy prices and less energy security - more individuals, communities and cities are looking to utilise autarkic principles to harvest, store and optimise use of local energy resources. Energy autarky can be described as *a location that relies on its own energy resources for generating the useful energy required to sustain the society within that region* or *a situation in which a region does not import substantial amounts of energy resources*. Functioning autarkic energy systems typically require a micro-grid, well understood energy demand and supply characteristics, opportunities for energy storage of various types and controls able to manage the harmonisation of system components. A critical additional ingredient is users who can work within constraints created by the adoption of autarkic principles. To elaborate the challenges and explore the issues involved with autarkic energy concepts this paper reports on the output from a workshop convened to investigate the role that energy autarky might play in delivering societies able to deliver the ambitious renewable generation targets set by both Scottish and UK Governments. In addition, monitored data from a community micro-grid system in Northern Scotland is analysed and presented to provide additional understanding of the complexities and opportunities created by an autarkic approach. The output from the workshop identified that whilst it is probable that a dogmatic interpretation

A.D. Peacock • E.H. Owens • S. Roaf (✉)
EGIS, Heriot Watt University, Edinburgh, Scotland, UK
e-mail: s.roaf@hw.ac.uk

D.W. Corne
School of Mathematical & Computer Sciences, Heriot Watt University,
Edinburgh, Scotland, UK

T.H. Karyono et al. (eds.), *Sustainable Building and Built Environments to Mitigate Climate Change in the Tropics*, DOI 10.1007/978-3-319-49601-6_7

of energy autarkic will not be universally applicable, the underlying principles represent a bottom-up way of widening participation in the development of future energy provision models. Whilst a number of issues and barriers were raised regarding its adoption, the attendees recognised that energy autarky represented a very positive and empowering vision for translating global scale issues to local energy transition. The analysis of monitored generation and demand data from a community micro-grid underlined the problems associated with supply-demand matching with intermittent generation and the need to place an emphasis on the community or entity as an open system that is able to participate in a full range of trading opportunities. Similarities were found between the types of behaviour necessary to create load responses relevant to energy networks containing large penetrations of renewable generation and communities set up based on energy autarky principles.

Keywords Autarkic • Energy • Low carbon living

1 Introduction

There are a rising set of socio-political circumstances and trends that are challenging the hegemony of incumbent energy provision models; prominent among these are energy security, climate change, resource depletion and a loss of faith in the capacity of centralised institutions, beholden to the primacy of shareholder value to provide democratised provision of energy. Allied to these factors is the rapid technological change being delivered in the fields of energy generation and communications that is creating economic opportunities for different visions of provision at radically different scales.

Scenario analysis conducted by the GB electricity transmission system operator forecasted that the proportion of installed generating capacity provided by embedded renewables would grow from a figure of circa 9% in 2010 to a figure of between 16% and 23% by 2030 depending on the scenario chosen (National Grid 2012). The realization of these forecasts, broadly seen as necessary if the UK is to meet its national targets and international obligations will require a change in the way power networks are configured, and controlled. The transition from the existing top down, predict and provide models to some form of distributed control approach has been widely predicted (Peacock and Owens 2014). This transition will necessitate a change in relationship between all actors in the energy supply chain; between the generator, the distributor, the retailer and the consumer with the demarcation between these groupings becoming increasingly blurred.

One emergent set of principles that may have the potential to provide a framework that addresses these socio-technical shifts is energy autarky. Energy autarky is variously described as a location that relies on its own energy resources for generating the useful energy required to sustain the society within that region (Müller et al. 2011) or a situation in which a region does not import substantial amounts of energy resources (Schmidt et al. 2012). The concept of energy autarky is not new but it has been gaining in prominence as a consequence of the factors listed above. For instance, 57 citations were found in Google Scholar for the term "energy autarky" before the year 2000, 98 between 2000 and 2010 and 214 already this decade. A number of issues abound with a dogmatic interpretation of energy autarky, for instance;

- how are boundaries defined and should embodied energy be included
- neo-classical trade theory suggests resultant reduction in trade would manifest itself as higher commodity and transaction costs
- security and resilience of energy supply may be impaired
- significant jeopardy of some regions experiencing transitional lock-out as a consequence of the approach not being universally applicable

These difficulties have led to a series of interpretations of the concept with variety further stimulated by it being an emergent paradigm that has not yet been subjected to efforts of standardization (Hansson 2010). For instance an entity may be described as energy neutral whereby it generates more energy locally than it consumes and trades the surplus (Rae and Bradley 2012); it may be described as net zero carbon (Sartori et al. 2012) that can be taken to mean that it produces surplus carbon free energy that it can trade to offset its consumption of carbon based energy; it can be described as stand-alone meaning that the energy it produces locally is sufficient to meet its needs although this definition can still permit the trade of forms of energy e.g. purchasing diesel for stand by generators.

A pragmatic version of energy autarky is used here describing the formation of an organising entity(s) whose role is to characterise energy demand for the local economy and services and to subsequently source this requirement from energy services derived from predominately locally renewable energy resources. In essence it will seek to create a supply condition that minimizes the use of imported fossil fuel based resources and maximize in its stead the probable use of local resources (Fig. 7.1). This definition effectively acknowledges the economic certainties that the imposition of a more dogmatic definition of energy autarky would result in a significant increase in costs and a likely diminution in service (Krajačić et al. 2011).

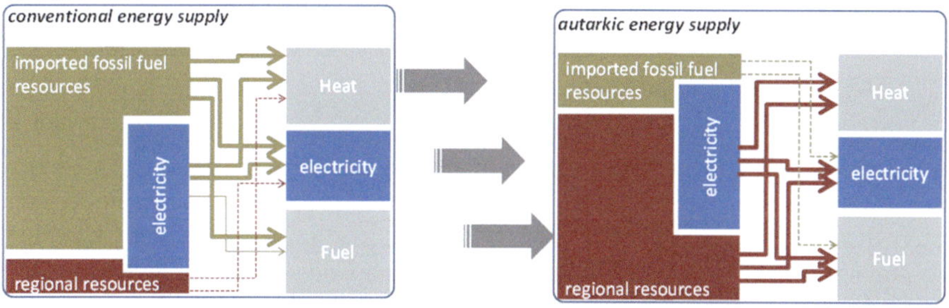

Fig. 7.1 Pragmatic autarky – transition to an autarkic energy supply motivated by a desire to reduce imported fossil fuel resources to a defined region or community

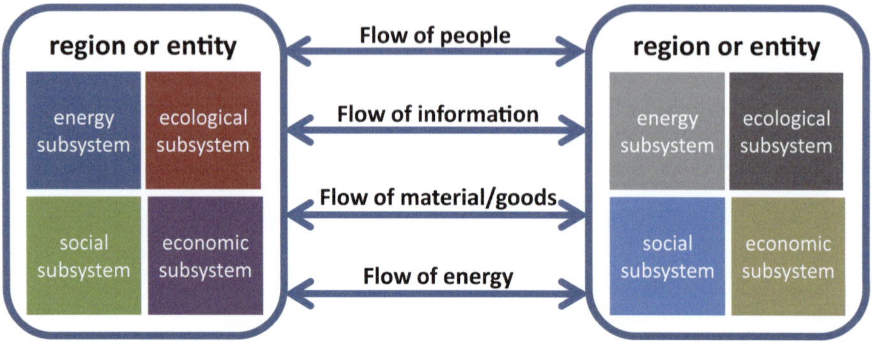

Fig. 7.2 Pragmatic autarky – regional boundaries and trading arrangements

Effectively this definition uses autarky to describe a destination that the entity may never reach but that will be used to determine its transitional direction.

The definition applied here places an emphasis on the entity as an open system that is capable of participating in a full range of trading opportunities (Fig. 7.2). This introduces the possibility of the entity creating value from managing energy demand and local supply in such a manner as to benefit wider societal efforts with respect to national and international obligations. Energy autarky as described then could represent a way in which participation in the ensuing energy transition can be broadened to encompass more actors than merely the owners of generating systems or the land on which they are deployed.

This chapter uses two approaches to explore autarkic concepts in more detail. In the first instance it reports on the output from a recent workshop involving energy practitioners and academics (ICARB 2014) that sought to describe social, economic and technical pathways that would allow Scotland to meet its governmental pledge to reduce 2050 carbon emissions by at least 80% from 1990 levels (Scottish Govt 2010). One pathway that was explored was the adoption and promotion of autarkic energy systems. In the second instance, monitored generation and demand data from the Findhorn Foundation Community was analysed to provide additional understanding of the complexities and opportunities created by an autarkic approach

2 Autarky Workshop

The key outcomes from the workshop are listed below but a core realization was that in effect many UK organisations are developing their own energy systems in ways that are in essence underpinned by autarkic principles. These included cities, communities, hospitals, schools, landed estates and land owning organisations such as the Forestry Commission as well as energy and water utilities. Through a series of interactive exercises based around the backcasting principal (Quist and Vergragt 2006) the following core findings regarding the energy autarky vision emerged:

- It was seen as being very positive and empowering for translating global scale issues to local energy transition
- It very firmly placed agency in the hands of the people creating more dynamic activity, local job creation and long term sustainability for communities
- It had the potential to provide substantial societal benefits, reduced sense of isolation and lack of connectivity and may help to bind communities through heightened social interaction
- Implementing the autarkic vision in urban settings where the majority of the population live would present greater challenges than in rural settings
- A number of economic issues would have to be addressed; namely issues about who would pay for grid connectivity, localised energy generation may actually increase costs, how start-up capital could be attracted and how autarkic approaches might have an impact on the wider economy and country wide GDP
- A range of issues were highlighted as being critical, namely; energy efficiency and retrofitting was key, development of system integration knowledge was critical, potential for DC grids, heat and power system should be championed, twinning urban areas with rural to allow renewable energy technologies to be correctly situated but allow urban areas to benefit economically; local autarky champions would have to be created.

The workshop participants recognised that the principal disruptive influence was likely to be cultural in dimension through the creation of new entrants into the energy market (e.g. local utilities, Consumer Co-ops, Housing Associations trading in energy). These new entrants may seek to exploit technology and communications advances to create change in energy provision to support a desire for more local/regional solutions.

3 ORIGIN Project

3.1 Introduction

In the ORIGIN project a sophisticated ICT system for the management of energy in a community has been developed and is being deployed based on the concepts of aligning energy demand with the availability of renewables-based supply, addressing community goals with respect to cost, carbon footprint and other objectives.

The hardware of the system comprises energy monitoring systems in buildings, actuators for selected devices, cloud based storage of data, and an associated communications infrastructure. At a temporal resolution of 30 min, the ORIGIN demand response architecture delivers the following; (a) 48 h ahead forecast of renewable generation, (b) 48 h forecast local demand, (c) quantified forecast of surplus generation, (d) identify a series of load shifting opportunities and (e) optimise the sequencing of these opportunities to produce a demand response schedule (Fig. 7.3). The overarching aim is to increase the proportion of renewable generation that is used

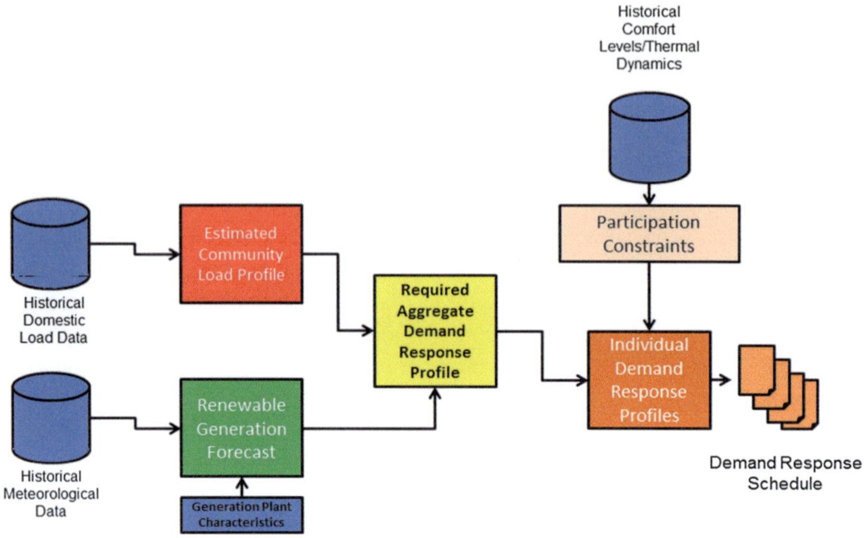

Fig. 7.3 ORIGIN system demand response architecture

Table 7.1 Description of key attributes of the Findhorn Foundation community energy systems

Location	Northern Scotland
Number of buildings	Circa 130
Installed solar-PV	25 kW
Installed wind generation	750 kW
Biomass district heating system	650 kW
Total installed capacity of air source heat pumps	75 kW
Total installed area of solar water heating systems	Circa 100 m^2
Electricity distribution network	Private wire; grid connected

locally and minimise the imports of energy to the community. Electricity, heat and transport energy vectors are included in the study which involves three communities; the Findhorn Foundation Community in Northern Scotland, the Damanhur community in Piedmont, Italy and the Tamera Healing Biotope in Southern Portugal.

To elaborate upon the challenges and explore the issues involved one of the three communities included in the ORIGIN project is considered, namely the Findhorn Foundation Community in Northern Scotland. The community owns and operates a mixture of renewable energy generation technologies and energy provision infrastructure including wind and solar-PV installations, a low voltage electricity distribution grid, an energy retailer and three distinct hydronic district heating systems fed by biomass boilers (Table 7.1).

3.2 Supply Demand Matching

The issues associated with achieving supply demand matching are illustrated by considering the production of renewable generated electricity from the community wind farm and the rooftop solar-PV systems in comparison with community electrical demand for the period 23rd to the 29th March 2014 (Fig. 7.4). The total demand for this period was 6640 kWh of which 30% was imported from the national grid despite the total renewable generation for the week being 21,843 kWh. Renewable electricity was generated almost entirely from the wind park with 79% being exported to the grid. This is not an atypical result and alludes to the issues associated with accommodating large capacities of renewable generation into existing grid infrastructure, designed over time to manage variability only with respect to load.

The issue of timing of export becomes critical and the results from a full calendar year reveal variation with time of day and with season, with winter, as expected, yielding higher capacity factors for wind generation and therefore higher levels of export from the community to the grid (Fig. 7.5). Peak export periods are during the afternoon, reflecting the predominant thermally induced wind regime in the UK. The value of the exported electricity to the operators of the national grid varies with time of day, currently in response to load variability, i.e. export is more valuable during peak demand periods than it is during the middle of the night. However, as the penetrations of renewable generation on the grid increase creating significant supply side variability, the value of export may become less associated with time of day and more associated with prevailing patterns of renewable availability.

A community that is able to manipulate its load profile in response to forecasts of renewable surplus may be creating a commodity whose value can be traded with the operators of the energy systems across different scales and geographical horizons. This illustrates the need to view autarkic systems as a series of connected entities that have the capacity to trade surpluses both between entities if plausible and with the incumbent centralised systems. The trading arrangements may yield the potential for communities that do not have access to either capital or to the land

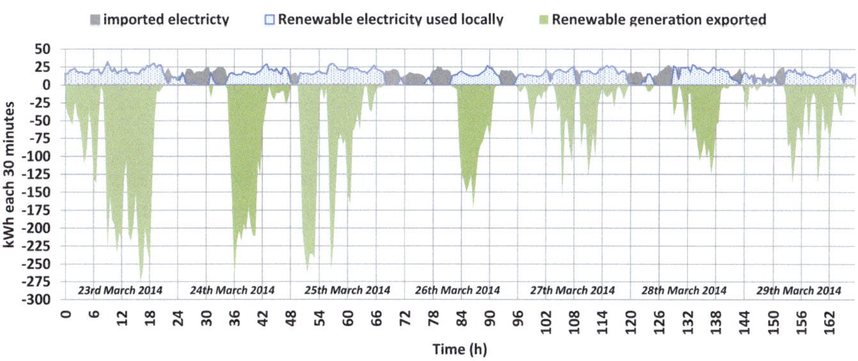

Fig. 7.4 Supply demand matching of renewable generation with local demand

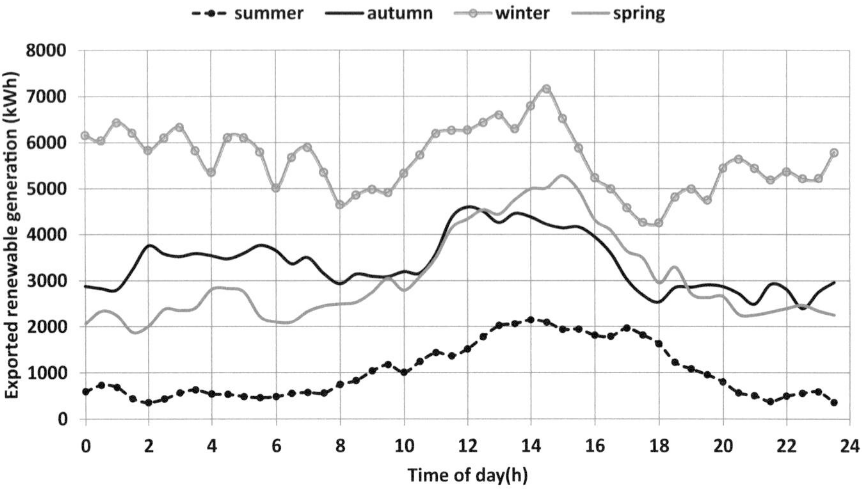

Fig. 7.5 Electrical export from the Findhorn Foundation Community to the national grid disaggregated by season and time of day

to participate in the ensuing energy transition through behavioural change to manipulate their load profile to help absorb supply side variability, democratizing participation and allowing wider societal aspects such as fuel poverty to be addressed.

Early results have shown that simply through the application of the user interfaces and the behavioural choices that have resulted from that access the homeowners in Findhorn are now actually importing around eighteen percent less grid energy than was usual before the systems were in place and in play. The results from Damanhur and Tamera are showing similarly significant drops in non-local energy use. As the systems are refined and the residents become habituated to them researchers are interested to learn if any falling off of the gains in local energy use result or any rebound effects become manifest in the data. A further area of interest will also be the actual cash savings that can be achieved by 'playing the system' with motivated homeowners.

3.3 *Further Work*

The ORIGIN project will seek to explore the boundaries of behavioural change both through informational systems and through direct actuation of consumer loads. With informational systems the forecasts of surplus will be provided to the community participants up to 48 h in advance and will be contextualised to make it relevant to the recipient. Electrical heating systems, buffered through thermal stores will be actuated directly based on the surplus forecast information to allow the resultant loads to better match to renewable generation. Overall the aim is to increase the proportion of locally generated renewable generation that is used locally.

4 Conclusions and Discussion

The achievement of true energy autarky is unlikely in all but the most specific of circumstances. However, its principles may become valuable in creating societal entities that can contribute locally towards the achievement of national and global energy policy aims. In the example provided, the Findhorn Foundation Community are participating in a project to explore ways in which loads can be manipulated via behavioural change and direct actuation to move the community closer towards autarkic principles. In so doing it is entirely feasible that they create a commodity, i.e. firm and predictable load profiles, that is of value to electricity network operators who are seeking to accommodate nationally driven capacities of renewable generation. The transition stimulated by the investigation of energy autarky by the community may alight upon behavioural change with respect to the timing of energy use and its relationship to intermittent renewable generation that may be of national and possibly global significance. The ORIGIN project will deploy the demand response software in autumn 2014 and will be reporting on the impact it has on the three communities from spring 2015. The results will be further interpreted in relation to both their Adaptation and their Mitigation impacts (www.origin-energy.eu).

Acknowledgements The authors would like to thanks the Findhorn Foundation for their participation in the ORIGIN project and for ICARB for the organisation and delivery of the workshop.

This paper is largely based on a paper given at the 2014 Asia Pacific Solar Research Conference in Sydney Australia with the conference presentation updates with the latest research outputs of the project.

References

Hansson SO (2010) Technology and the notion of sustainability. Technol Soc 32(4):274–279

ICarb (2014) http://icarb.org/2014/06/02/backcasting-summit. Accessed Sept 2014

Krajačić G, Duić N, Zmijarević Z, Mathiesen BV, Vučinić AA, da Graça Carvalho M (2011) Planning for a 100% independent energy system based on smart energy storage for integration of renewables and CO2 emissions reduction. Appl Therm Eng 31(13):2073–2083

Müller MO, Stämpfli A, Dold U, Hammer T (2011) Energy autarky: a conceptual framework for sustainable regional development. Energy Policy 39(10):5800–5810

National Grid (2012) Electricity ten year statement. Origin project final reports are available on: www.origin-energy.eu

Peacock AD, Owens EH (2014) Assessing the potential of residential demand response systems to assist in the integration of local renewable energy generation. Energy Effic 7(3):547–558

Quist J, Vergragt P (2006) Past and future of backcasting: the shift to stakeholder participation and a proposal for a methodological framework. Futures *38*(9):1027–1045

Rae C, Bradley F (2012) Energy autonomy in sustainable communities—a review of key issues. Renew Sustain Energy Rev 16(9):6497–6506

Sartori I, Napolitano A, Voss K (2012) Net zero energy buildings: a consistent definition framework. Energy Build 48:220–232

Schmidt J, Schönhart M, Biberacher M, Guggenberger T, Hausl S, Kalt G, Schmid E (2012) Regional energy autarky: potentials, costs and consequences for an Austrian region. Energy Policy 47:211–221

Scottish Govt, (2010), http://www.scotland.gov.uk/News/Releases/2010/09/23134359. Accessed Sept 2014

Part II
Adapting City for Climate Change

Chapter 8
Adapting City for Frequent Floods: A Case Study of Jakarta, Indonesia

Tri Harso Karyono and Firmansyah Bachtiar

Abstract Due to global warming and climate change, floods have occurred quite often in many places in the world, including Jakarta. As the capital of Indonesia, Jakarta has suffered from floods since a long time ago, even before the Dutch colonized this country. The recent floods in this capital are likely caused by more complex factors compared to the previous ones. To respond to such complexity factors which have created floods in this capital, an urban design strategy is proposed. The design strategy proposes that parts of the city, particularly those which are affected by frequent flooding are to be elevated. Most of the city centre is heavily affected when floods occur, and these areas are where the multi-storey and high rise buildings are built. By elevating the main roads in these areas and connecting these roads to the upper floors of the buildings, people can gain access easily to almost all of the buildings from the elevated roads. The ground surface areas of the city can then be redesigned in such a way as to increase their ability to absorb more rain water and overflowing river water, reducing the flood effects in the city.

Keywords Climate change • Elevated city • Frequent floods • Global warming • River water • Rain water • Sea water

1 Introduction

Jakarta, located at 6° 12′ south latitude and 106° 48′ east longitude, is the capital of Indonesia. The area of Greater Jakarta or the Jakarta Special District is 662 km^2 of land area and 6977 km^2 of sea area (Jakarta.go.id 2016). The old Jakarta was originally formed by a small town called Jayakarta and later Batavia during the Dutch occupation. Batavia was the official name of the capital city of the Netherlands Indies in the nineteenth century, which is now the district is called Kota Tua (old town) in the northern part of modern Jakarta (Merrillees 2000).

T.H. Karyono (✉) • F. Bachtiar
School of Architecture, Tanri Abeng University, Jakarta, Indonesia
e-mail: t_karyono@yahoo.com

T.H. Karyono et al. (eds.), *Sustainable Building and Built Environments to Mitigate Climate Change in the Tropics*, DOI 10.1007/978-3-319-49601-6_8

Situated in in the warm and humid tropical climate, Jakarta has two main seasons: the dry and rainy seasons. There is almost no temperature difference throughout the year. The monthly average outdoor temperature and daily temperature are about the same throughout the year, which is around 28 °C with the range of temperature between 24 and 34 °C in the dry season, which takes place between April and September and about 23 and 32 °C in the rainy season which takes place between October and March. During the rainy season, the outdoor temperature tends to be slightly lower by about 1–2 °C than in the dry season (Climate of Indonesia 2016). The average rainfall of Jakarta per annum is around 2000–2500 mm (BPS-Indonesia 2013).

The history of flooding in Jakarta has been reported since the Tarumanagara Kingdom in the fifth century. The Kingdom was located not far from the site of modern Jakarta today. The Tugu inscription, written in Sanskrit on a one meter oval stone found in Tugu village, North Jakarta in the early fifth century, described a number of hydraulic projects; the irrigation and water drainage project and the digging project to straighten and widen the river. These projects were conducted in order to avoid flooding in the wet season, and to provide irrigation during the dry season (Sudaryat 2014). When the Governor General Jan Pieterszoon Coen, the 4th Governor-General of the Dutch East Indies (Indonesia) started his office, he began to build ports and canals in Batavia in order to try to solve the city's frequent flood problems. The major floods which have occurred in Jakarta have been recorded by the Dutch many times since 1621. Recent floods were recorded in 1976, 1984, 1994, 1996, 1997, 1999, 2002, 2007, 2008, 2013, and 2014 (Caljouw et al. 2004; Anggara 2014).

Looking at the time span in which the floods occurred, it is clear that flooding in Jakarta has occurred more often within recent time. This gives a hint that Jakarta will experience more problems in the very near future due to the incoming floods in this city. Global warming and climate change could be the main factors causing these more frequent floods in this city.

2 Climate Change and Frequent Floods

Climate change affects the changes of many things on earth. The unpredictable climate and weather create a number of problems. These problems are not only for farmers and fishermen, the problems have been affecting the whole population living all over the world, in the villages, towns and big cities. Some areas become very dry due to the absence of rain and others have too much precipitation, creating floods, even in the dry season. Due to climate change, millions of people worldwide livings in coastal cities like Jakarta suffer from frequent floods which occur at any time regardless of the seasons.

Along with the rise of the sea level, Jakarta has suffered from floods because of its massive development in which many of the green areas have been sacrificed and converted into buildings, roads, car parks and other facilities which have covered

Fig. 8.1 Jakarta with frequent floods occurred mostly every year (Source: Warsito)

the permeable surface lands. Consequently, this has reduced the ability of ground surfaces to absorb the rainwater. Moreover, the lack of an integrated drainage system makes the city more vulnerable to the floods. In order to respond to the frequent floods, either caused by the rising seawater or by heavy rains, strategies have to be developed to adapt the city to the floods. Frequent floods in Jakarta which take place annually during the rainy season have created a serious problem in this capital city (Fig. 8.1). A number of recent floods in Jakarta would not have had such a severe impact had the city's authorities anticipated the disaster earlier.

There are three main aspects associated with the causes of flooding in Jakarta, namely local, regional and global aspects (Karyono 2013). The first cause of Jakarta's flooding is the local factor. Many urban experts argue that the lack of open spaces and green areas in Jakarta is a source of the floods, as rainwater cannot be absorbed by the ground surfaces. This argument could be partly true as many cases show the decrease of the green areas in a number of cities tends to be correlated with the increase in number of flooding events in a number of Indonesian cities. However, looking further about the reason why more Indonesian cities are flooded during the rainy season, the lack of integrated drainage systems could be the main factor. The ground surfaces of any city tend to be covered primarily by hard materials in the form of buildings, roads, parking lots and other amenities. Most cities in the world would have a similar situation in which more of their surfaces are covered with materials that promote rainwater run-off and only a small portion would be absorbent materials. When there is a campaign of greening the city, the idea might not be solely about improving the capability of the city to absorb more rainwater, but in addition to improve the O_2 production and to reduce carbon emissions, as well as to reduce outdoor temperatures. More vegetation planting in the city would reduce the urban heat island (UHI) effect of the city. With a lower outdoor temperature, a tropical city can reduce its energy consumption for cooling. Other than fulfilling the aesthetic aspect, greening the city will also help to absorb carbon dioxide and dust. Plants need carbon dioxide to grow.

In a green city concept, particularly in the Tropics, vegetation can be planted anywhere. Plants do not need to be planted on the ground only; green areas can also be established on rooftops and on the building's walls and balconies with potted plants. The main idea of greening is to minimize the amount of hard material being exposed to the sun, so that the penetration of solar radiation can be kept to a minimum. In the tropical cities, such as Jakarta, with heavy rainfall which comes very often, it is not reasonable to expect the rainwater to be fully absorbed by the ground surfaces.

A large number of cities across the globe are dominated by hard ground surfaces, such as concrete and asphalt but they remain free of flooding despite heavy rains. Examples include Hong Kong and Singapore. The rapid growth of Jakarta has replaced many green spaces by buildings, parking lots and roads. Rainwater that falls on the hard surfaces must be able to flow through water channels to lower destinations like canals or rivers, eventually ending up flowing to the sea. This process can be done systematically by means of an integrated drainage system. Unfortunately, Jakarta does not have an integrated drainage system and even lacks appropriate water channels.

Old channels, built a decade after the country proclaimed its independence from the Dutch, are still used today. Unfortunately, they are no longer sufficient to accommodate the current rainwater levels. Rapid urbanization has led to more extensive parts of the city being covered by hard surface materials, while human activities have also led to the serious obstruction and deterioration of the city's drainage system, reducing its discharge capacity.

Since the ground surfaces are now almost entirely covered by concrete and asphalt, the runoff rainwater becomes many times greater than before where there were more absorbent surfaces and, consequently, inundates roads, open spaces and some single-story houses in the city.

Constructing an integrated drainage system would reduce, if not totally overcome, Jakarta's floods. Although the entire surface of the city is covered by concrete and asphalt, the city will not be flooded if it is well equipped with an integrated drainage system. All the runoff water would be collected and circulated via the water channels and directed to the sea. The dimension of the channels is also an important consideration. The water channels should be sufficiently large to accommodate the runoff water and so avoid overflowing water that causes flooding. Beside an integrated drainage system and an increase in the size of water channels, there needs to be an ongoing program to dredge sediment from the city's rivers, canals and lakes. The ability of these urban waterways to contain runoff rainwater has markedly declined in recent years due to sedimentation.

The second aspect associated with the cause of flooding in Jakarta is the regional factor. From the regional point of view, Jakarta's flooding is the result of large amounts of water from upstream rivers in West Java that flow across the capital. Deforestation in the headwater regions caused by the development of farms, plantations and also new housing has reduced the upstream rainwater absorption, sparking massive water flows to lower areas and, hence, creating huge flooding in Jakarta.

The government has a plan to develop artificial lakes to act as water reservoirs along the West Java border outside Jakarta to avoid the floods. This plan aims to divert the overflowing river water before reaching Jakarta. Previosly the West Java authority was reluctant to provide any land to support this purpose, however, under the new government, this plan is ready to be implemented shortly.

The third aspect, viewed as the main aspect which creates frequent floods in Jakarta, is the global aspect. In the global aspect, all coastal cities in the world, including Jakarta, face a high potential of flooding due to global warming. More than half a billion people worldwide living on small islands and in coastal cities are faced with a serious threat of flooding as the sea level continues to rise due to the melting of polar ice due to global warming (Girardet and Mendonca 2009).

Over the past 3000 years the sea levels had risen at the rate of 0.1–0.2 mm per year, but in the twentieth century, due to global warming, the sea level rises have increased to 1–2 mm per year, or about ten times faster than before (Roaf 2009). This phenomenon creates some heavy flooding, particularly in the coastal towns. The coastal areas and cities like Jakarta could have flooded even in the absence of rain as the risen seawater enters the land.

Global warming, in which the average earth temperature rises, is caused by excessive greenhouse gas (GHG) emissions. Approximately 78 % of greenhouse gas GHG emissions in the atmosphere are in the form of CO_2, the rest is methane (CH_4), nitrous oxide (N_2O), sulphur hexafluoride (SF_6), water vapour and ozone (IPCC 2015). Data from 2009 showed that approximately 30.4 billion tonnes of carbon were released into the atmosphere (Rogers and Evans 2011)). By 2009 global CO_2 concentration had reached 387 parts per million (ppm), an increase of about 40 %, from 275ppm in 1900 (Girardet and Mendonca 2009). Buildings account for about 15 % of emissions, while transport and industry were 14 and 21 %, respectively, and the remainder was emitted by other activities (Karyono (2011).

To reduce global warming, the Kyoto Protocol (United Nations 1998, Anderson 1998) suggested that a number of industrial countries would reduce their carbon emissions by 5 % from 1990 levels by the years of 2008 until 2012. In an evaluation of the impact of the Kyoto protocol on reducing global carbon emissions after 2012 (Böhringer n.d; Morel and Shishlov 2014)it was recorded that some of the developed countries like UK, France and Germany, and most of Eastern Europe are successfully reducing by more than the targets, while other industrialized countries like Canada, Australia and Japan are failing to reach the mandated goals (Clark 2014).

On average, the developed countries, excluding the US which refused to sign the treaty, have slightly reduced their carbon emissions between 2008 and 2012. However, many developing countries, such as China, with no obligation to reduce emissions, have significantly increased their carbon emissions (Clark 2014). After all this, global carbon emissions are still increasing after the Kyoto Protocol, reaching up to 34 billion tonnes recently, threatening a global temperature rise up to 7 °C in 2100 (Oliver et al. 2012). The Doha climate change conference on November 2012 extended the Kyoto Protocol to reduce global carbon emissions, which would take effect until 2020 (Ritter and Casey 2013).

However, even if the Doha commitment is successfully achieved, the fate of global carbon emissions is not covered by it. Many poor and developing countries have a right to develop to improve their people's wellbeing and reduce their poverty. Improving people's wellbeing would enhance the use of energy, and unfortunately would release much more carbon dioxide. While the developed countries are trying to reduce the global carbon emissions, the developing countries would be avoidably increasing the global emission of carbon dioxide.

In the current view, it seems there is not much choice to avoid global warming in the near future. The sea level will rise steadily, creating more floods in the coastal cities like Jakarta. There is a need to adapt to global phenomena to reduce the impact of frequent floods occurring in Jakarta due to the rise of the sea level.

3 Solving the Problems: Raising the City

Considering the rise of the sea level due to global warming, Jakarta would have more frequent floods in the future. There is a need to provide a future design scenario for the city. Looking back to the past in which digging up the city to make canals was not successfully saving Jakarta from the floods, it is time to find a different approach to solving Jakarta's flooding. It is worth learning what our ancestors have done in the past to manage a sustainable environment. Our ancestors wisely managed their immediate environment in such a way as that they could preserve it for such a long time without being damaged. In a number of vernacular societies, in Indonesia and elsewhere, people build stilt houses to protect them mainly from their enemies and from wild animals like tigers. Besides these objectives, stilt houses are good for avoiding floods. Since the house is lifted up above the ground, the ground surface beneath the house is open, uncovered, allowing it to absorb the rainwater, thus helping to prevent the floods.

If we want to create a free-flood city, a city which has no floods, we need to look back to what our ancestors did. Perhaps we may not need to copy our ancestors' way of life or their creations, but we can learn from them and use them to solve our current problems. Using the idea of the way a stilt house could respond to floods, we propose to build the city of Jakarta up from the ground. The city may be built on stilts, in which its multi storey buildings and roads can be lifted up from the ground, leaving the ground surface open (Karyono 2014). The open ground surface of the buildings' ground floors is built with non-absorbent materials, such as concrete blocks, so they do not easily absorb the rainwater. In this proposal, they have to be removed, replaced by water absorbent material and surfaces. In case of flood, this surface would help to absorb the water. However, this space is not allowed to be used for permanent activities, except for vertical movement purposes, such as stairs, escalator and elevator. No basement is allowed to be built in this area and those which are already built must be used as a water tank to accommodate the runoff water.

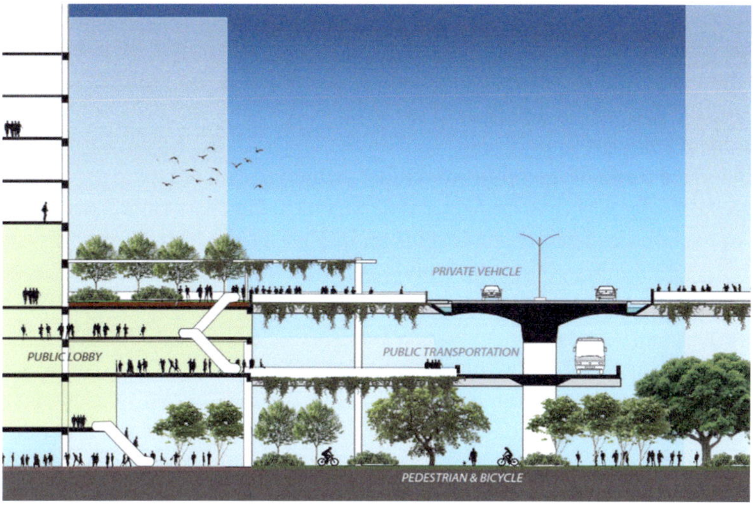

Fig. 8.2 Raising the city to avoid frequent floods

The opened ground floors of the buildings are not only beneficial to absorb the excessive rainwater throughout its ground surfaces, but also provide space to accommodate this excessive rainwater in case of flooding. When this idea is applied to up 60 % of the existing multi storey buildings and to all the new buildings, the size of uncovered ground surfaces of Jakarta would increase significantly and would reduce the runoff water greatly.

Providing open ground floors for many multi-storey buildings in the town centre without walls will enhance air movement around the buildings. This would create a cooling effect in the outdoor spaces in a warm tropical city such as Jakarta. Furthermore, in the very near future buildings must not be fenced. Any outdoor spaces surrounding the buildings at city centre should be considered as public spaces. Everybody can use them. Hence, there will be many possibilities for short-cut access in the city as all the areas surrounding the buildings are now public spaces. People could move around the city by walking and cycling around and through the ground floor of the buildings.

Main roads are lifted up as they are to be used for motorized vehicles, while beneath these roads, there are small roads for cycling, pedestrian pathways, and gardens. There will be a division between unmotorized vehicles and motorized vehicles in entering the building. The pedestrians and cyclists will use the space on the ground levels and enter the building from the ground floor entrance, while people using motorized vehicles, mainly public transport like monorail, will enter the buildings via the lobbies at the upper floors of the buildings, such as at the second or third floors at the same level as the flying roads (Fig. 8.2).

The question is how to materialize this idea. How to remove the ground floors of many existing buildings in Jakarta to be used as rain water absorbers? How to change the use of basements from mainly a car park facility into a rainwater tank in case of flooding? The Jakarta authority will have their duty to issue such a decree to mandate the buildings' owners in the flood-risk area to alter their buildings. To avoid the disruption of activities inside and surrounding the buildings, this can be done step by step within a particular period.

The local authority should also build all the upper roads or fly-over roads step by step to replace some of the on-ground roads, which in turn are to be used mainly for the pedestrians and cyclists. Implementing this idea, the city would have at least two levels or even more. The ground level is mainly used to absorb the runoff water and is used for non motorized vehicle activities, while the upper ground level would be used to accommodate mainly the motorized vehicles and in case of flood this can be used for all.

Lifting the city up by leaving the buildings' ground floors uncovered so as to provide more uncovered ground surfaces is like the idea of stilt houses of the vernacular society, and to provide flying roads around the city is a way to adapt the city to global warming and climate change. When it is implemented, Jakarta would reduce its flood risks.

References

Anderson JW (1998) The Kyoto protocol on climate change – Background, Unresolved issues and next steps, available at http://www.rff.org/files/sharepoint/WorkImages/Download/RFF-RPT-kyotoprot.pdf. Accessed 18 Feb 2016

Anggara (2014) History of Jakarta flood: from Purnawarman to Joko Widodo, Jakarta by Train, 25/01/2014, available at http://jakartabytrain.com/history-of-jakarta-flood-from-purnawarman-to-joko-widodo/. Accessed 17 Feb 2016

Böhringer C (n.d.) The Kyoto protocol: a review and perspectives, Discussion paper No. 03-61, Centre for European Economic Research, available at https://ub-madoc.bib.uni-mannheim.de/137/1/ZEW26.pdf. Accessed 18 Feb 2016

BPS-Indonesia Statistical Beareu, *Rainfall* (2000–2013) http://www.bps.go.id/linkTabelStatis/view/id/1349. Accessed 17 Feb 2016

Caljouw M, Nas PJM, Pratiwo (2004) Flooding in Jakarta, Paper in the *first international urban conference*, Majapahit Hotel, Surabaya, Indonesia, August 23 – 25, 2004, National Capital Integrated Coastal Development

Clark D (2014) Has the Kyoto protocol made any difference to carbon emissions?. The Guardian Environmental Blog, 26 November 2012, available at http://www.theguardian.com/environment/blog/2012/nov/26/kyoto-protocol-carbon-emissions. Accessed 18 Nov 2014

Climate of Indonesia (2016) Available at https://en.wikipedia.org/wiki/Climate_of_Indonesia. Accessed 17 Feb 2016

Girardet H, Mendonca M (2009) A world renewable energy, ecology, equality a report for the world future council. Gren Books Ltd, Totnes

IPCC (2015) Climate change 2014 – Sysnthesis report, Intergovernmental panel on climate change (IPCC), Geneva, Switzerland, available at https://www.ipcc.ch/pdf/assessment-report/ar5/syr/SYR_AR5_FINAL_full.pdf. Accessed 18 Feb 2016

Jakarta.go.id (2016) Gografis Jakarta available at http://www.jakarta.go.id/v2/news/category/geografis-jakarta. Accessed 17 Feb 2016

Karyono TH (2011) Global warming and green architecture. Jakarta Post 5 November 2011

Karyono TH (2013) Frequent cases of flooding reduces Jakarta to third-world city. Jakarta Post 10 February 2013

Karyono TH (2014) Floods and raising the city. Jakarta Post 5 April 2014

Merrillees S (2000) Batavia in nineteenth century photographs. Archipelago Press, Richmond

Morel R, Shishlov I (2014) Ex-post evaluation of the Kyoto protocol: four key lessons for the 2015 Paris agreement, Climate Report – Research on the economics of climate change, No 44, May 2014, available at http://www.cdcclimat.com/IMG/pdf/14-05_climate_report_no44_-_analysis_of_the_kp-2.pdf. Accessed 18 Feb 2016

Olivier JGJ, Janssens-Maenhout G, Peters JAH (2012) Trends in global CO_2 emissions 2012 report, PBL Netherlands Environmental Assessment Agency, The Hague, available at http://edgar.jrc.ec.europa.eu/CO2REPORT2012.pdf. Accessed 18 Feb 2016

Ritter K, Casey M (2013) UN climate conference: Kyoto protocol extended at Doha, Qatar Talks, *Huff Post Green*, 7 February 2013, available at http://en.wikipedia.org/wiki/Global_warming. Accessed 8 Feb 2016

Roaf S (2009) Adapting buildings and cities for climate change survival guide. Routledge, Amsterdam

Rogers S, Evans L (2011) World carbon dioxide emissions data by country: China speeds ahead of the rest, *The Guardian*, 31 January 2011, available at http://www.theguardian.com/news/datablog/2011/jan/31/world-carbon-dioxide-emissions-country-data-co2. Accessed 18 Feb 2016

Sudaryat Y (2014) The structure of ancient sundanese language, TAWARIKH: International Journal for Historical Studies 6(1), October 2014, Bandung, Indonesia

United Nations (1998) Kyoto protocol to the United Nations framework convention on climate change, available at http://unfccc.int/resource/docs/convkp/kpeng.pdf. Accessed 18 Feb 2016

Chapter 9
STEVE Tool Plug-in for SketchUp: A User-Friendly Microclimatic Mapping Tool for Estate Development

Steve Kardinal Jusuf, Marcel Ignatius, Nyuk Hien Wong, and Erna Tan

Abstract Due to increasing urbanization and heat island issues, current cities need to sensibly plan their future developments to preserve the quality of their urban environment. An urban climatic map (UCmap) method can be utilized for this purpose to examine the effectiveness of various urban environmental solutions in improving environmental conditions, particularly in terms of outdoor temperature and human thermal comfort. The available climatic models for both temperature and thermal comfort mapping can be deemed complicated and less user friendly; particularly for practitioners such as urban planners. Hence, the Screening Tool for Estate Environment Evaluation (STEVE Tool) and Thermal Sensation Vote (TSV) models were developed with a motivation to link research findings with urban planners. This paper showcases the models which have been developed as a plug-in for Trimble SketchUp (formerly Google SketchUp and hereinafter simply SketchUp); serving as a straightforward analysis tool for urban planners during the design and feasibility study processes. This plug-in can generate various climatic maps and sectional temperature profiles. Furthermore, it is also equipped with a plants database from the National Parks Board (NParks) and thus is able to calculate the impact of landscaping on temperature and carbon sequestration accordingly. A case study has been provided, using a central business district (CBD) area in Singapore. In the end, the generated UCMaps are able to provide microclimatc information for an existing masterplan or urban morphology layout, so that planners or architects can identify the hot spots or problematic areas and then solutions or mitigation strategies can be developed afterwards by analysing the UCMaps.

Keywords Urban climatic map • Microclimate • Tropics • Outdoor temperature • Prediction models • Outdoor thermal comfort

---

S.K. Jusuf (✉)
Sustainable Infrastructural Engineering (Building Services), Singapore Institute of Technology, Singapore, Singapore
e-mail: stevekj@singaporetech.edu.sg

M. Ignatius • N.H. Wong • E. Tan
Department of Building, School of Design and Environment, National University of Singapore, Buona Vista, Singapore

T.H. Karyono et al. (eds.), *Sustainable Building and Built Environments to Mitigate Climate Change in the Tropics*, DOI 10.1007/978-3-319-49601-6_9

1 Introduction

Urban climate is one of the elements of urban physical environments, which is often ignored in urban planning. To design a sustainable city, it is necessary to factor the climatic information holistically and strategically into the planning process. Both macro and microclimatic analysis have become essential due to rapid urbanization, higher density, and more compact cities; hence the concept of the urban climatic map (UCmap) was developed to understand the climate phenomenon in the urban areas.

Since the 1970s, UCMap has been developed by German researchers, which have a strong focus on applied urban climatology (Matzarakis 2005). It is considered as an appropriate tool for translating climatic phenomenon and problem into 2-D images and symbols with land use and spatial information for the urban planning use.

It can help urban planners, architects and governors to understand and evaluate the effect of urban climatic issues on decision-making and environment control. With the development of UCMap since the 1990s, it has become a worldwide research interest in which Germany, Hong Kong, and Japan have been the exemplars in conducting urban climate analysis and application (Katzschner 1988; Katzschner and Mülder 2008; Ng et al. 2009; Ren et al. 2011; Tanaka and Moriyama 2011).

One important focus of research under the urban climate scope is the urban heat island (UHI) phenomenon. UHI has been widely known as the result of city growth due to urbanization, which brings consequences such as local climate alteration and adverse impacts on the outdoor environmental quality (Lo and Quattrochi 2003; Rohington 2005; Taha et al. 2000).

This paper showcases how to use the UCmap concept integrated with the established prediction models to analyse outdoor temperature and thermal comfort, the aspects that are highly affected by the UHI. The prediction models, STEVE Tool and TSV, have been developed using the Singapore climate as the model of tropical high density urban areas. The purpose is to link research findings, particularly in air temperature prediction models, with urban planners. The prediction models were further developed and integrated as a plug-in into SketchUp for use as an analysis tool for urban planners in the planning process.

2 Temperature Prediction Tool for Estate Development

Many urban climatic studies have been carried out in identifying and mitigating the UHI effect. A statistical model was developed in Szeged, Hungary to determine the relationship between UHI intensity and urban surface parameters, such as sky view factor (SVF), building height, built-up ratio and water surface ratio (Bottyán and Unger 2003). In Hong Kong, critical variables that can mitigate both daytime and nocturnal UHI are found to be SVF, surface albedo, altitude, and vegetation above 1 m in height (Giridharana et al. 2007). The temperature mapping method can be utilized to examine the effectiveness of the various urban environmental solutions such as greenery, cool roof and facades, etc. to improve the environmental conditions, comfort and consequently the energy efficiency of buildings (Alfaro et al. 2006).

In the context of Singapore, as a tropical urban role model, the changes of outdoor temperature depend on the urban morphology (Wong et al. 2011), and its pattern is closely related to urban land use. In general, parameters such as amount of greenery, height of buildings and width of urban or street corridors have been discovered to have direct impact on urban air temperature. Interactions between these urban surfaces and the ambient conditions form the basis of urban climate models that have been used to calculate the microclimate air temperature in an urban estate.

An air temperature prediction tool for the Singapore climate called the STEVE Tool has been developed, which utilizes the correlation parameters between building – pavement – greenery and ambient temperature. It was developed based on empirical data collected over a period of time as part of the development of an assessment method to evaluate the impact of estate development. The empirical model calculates the air temperature of a point of interest based on a 50 m radius in an urban built up area in Singapore (Jusuf and Wong 2009, 2016; Wong and Jusuf 2008; Wong et al. 2007, 2011). The STEVE tool is used to simulate the ambient temperature condition with wind consideration, surface customization, and proposed greenery options.

The air temperature prediction models can calculate the daily minimum (T_{min}), average (T_{avg}) and maximum (T_{max}) temperature of each point of measurement based on climate predictors and urban morphology predictors. The climate predictors are daily minimum temperature (T_{min-r}) at reference point, daily average (T_{avg-r}) temperature at reference point, maximum (T_{max-r}) temperature at reference point, average of daily solar radiation total (SOLAR*total*), and average of solar radiation maximum (SOLAR*max*); while the urban morphology predictors are percentage of pavement area over surface area (PAVE), average height to building area ratio (HBDG), total wall surface area (WALL), Green Plot Ratio (GnPR), SVF and average surface albedo (ALB) (Wong et al. 2011).

The model was redeveloped in 2012 with additional field measurements in a wider area of Singapore in order to be able to predict ambient temperature on various land uses and urban layouts. Overall, weather sensors and loggers were installed at 10 different areas representing land uses such as business parks, industrial estates, CBD, parks, and housing estates (see Fig. 9.1 and Table 9.1). This amalgamates the data on the microclimate and urban morphological parameters gathered from both the current and past field measurements. In remodeling of the air temperature prediction models, past air temperature measurement and geographical information data tabulated by Jusuf and Wong (2009) were combined with recent data collected from the new locations.

It is important to note that during model redevelopment, only days with fairly clear, calm and hot weather conditions were selected for analysis; mainly because

Fig. 9.1 Field measurement conducted in various locations to support the STEVE tool development

Table 9.1 Field measurement locations and number of measurement points for each area

No	Area	Number of measurement points
1	One-North Business Park	30
2	International Business Park	10
3	Loyang Industrial Estate	11
4	Woodlands Industrial Park	12
5	Marina Bay Promenade and Promontory	8
6	Gardens by the Bay	4
7	Shenton Way	7
8	Tanjong Pagar	7
9	Tampines	4
10	Bedok	12
	Total	**105**

high UHI intensity occurs in those conditions (Kolokotroni et al. 2006). The selection of days was done by analyzing the profile of daily solar radiation, which exhibits a bell curved shape with a peak exceeding 700 W/m^2 and using days with wind speed less than 3 m/s. Although the field measurements were conducted during different periods of time, by selecting data only from days with clear, calm and hot weather conditions, the data were comparable for analysis (Fig. 9.2). In addition, urban morphology parameters from the built up areas were also compiled, e.g. SVF, GnPR, HBDG, WALL, and PAVE.

The field measurements conducted in this research used the following instrumentation to collect relevant temperature, wind, and sky view factor (SVF) variables (Table 9.2):

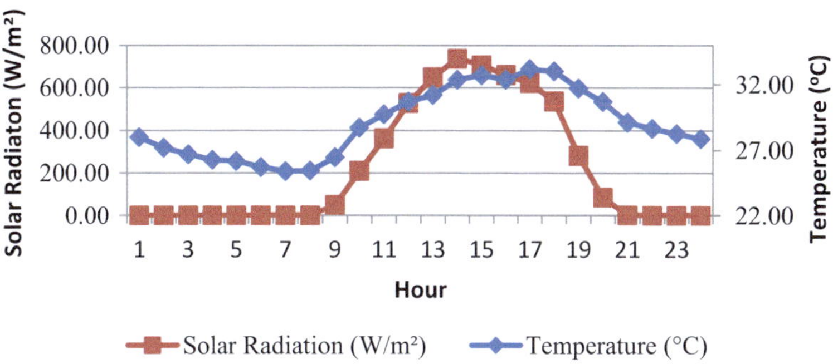

Fig. 9.2 Solar radiance and temperature profile for a typical day selected

Table 9.2 Field measurement instrumentations

Temperature/Relative humidity	Wind speed	Sky view factor
Equipment: *Hobo U12–011*	Equipment: *S-WSA-M003*	Equipment: *Nikon D80 Digital camera and Nikon AF DX 10.5 F2.8G IF ED fisheye lens*
Installed at the height of 2.1 m	Installed at the height of 2.1 m	Lensconstruction (groups/element): 5/8;
Temperature measurement range: −20–+70°C;	Range: 0–45 m/s;	Picture angle with 35 mm (135) format: 180°;
Temperature accuracy: ±0.35° from 0–50°C;	Operating temperature range: −40 °C–75 °C;	Minimum f/stop: 22;
RH measurement range: 5 – 95 % RH non-condensing;	Accuracy: ±1.1 m/s or ±4 % of reading whichever is greater;	Closest focusing distance: 0.25 m;
RH accuracy: +2.5 % RH from 10 % to 90 % RH (typical)	Resolution: 0.38 m/s;	Maximumreproduction ratio: 1/10.
	Starting threshold: 1 m/s;	
	Maximum wind speed survival: 54 m/s.	

Simultaneous multiple regressions were conducted to determine the best linear combinations of the independent variables. Regressions analyses were conducted for empirical T_{min}, T_{avg} and T_{max} models based on the factors that affect the urban ambient temperature. To get better understanding on the air temperature during daytime and nighttime, the daily average temperature for daytime at 07.00–18.59 h ($T_{avg(daytime)}$) and for nighttime at 19.00–06.59 h ($T_{avg(nighttime)}$) were modelled as well. With this diurnal segregation, any change in the correlation behavior due to the influence of solar radiation on air temperature can be detected. Furthermore, the redeveloped models incorporated the influence of wind speed ($WIND_{max}$) on the air temperature and to enhance the model's prediction capability. The models are as shown in the following Eqs. 9.1, 9.2, 9.3, 9.4 and 9.5:

$$T_{max}\left(°C\right) = 4.494 + 0.695\, Ref\, T_{max}\left(°C\right) + 0.001\, SOLARmax\left(W/m^2\right) - 0.39WINDmax + 0.008PAVE\left(\%\right) - 0.002 Avg\, Height\left(m\right) + 1.078 SVF + 32.235 ALB \quad (9.1)$$

$$T_{avg}\left(°C\right) = 2.676 + 0.919\, Ref\, T_{avg}\left(°C\right) - 0.201 GnPR - 0.015 HBDG + 1.520E - 05 WALL\left(m^2\right) + 0.240 SVF \quad (9.2)$$

$$T_{avg(daytime)}\left(°C\right) = 7.619 + 0.733\, Ref\, T_{avg-day}\left(°C\right) + 3.61E10 - 5 SOLAR_{total}\left(W/m^2\right) + 0.004 PAVE\left(\%\right) - 0.013 HBDG + 0.669 SVF \quad (9.3)$$

$$T_{avg(nighttime)}\left(°C\right) = 2.347 + 0.904 RefT_{avg-night}\left(°C\right) + 5.786E - 05\, SOLAR_{total}\left(W/m^2\right) + 0.007 PAVE\left(\%\right) - 0.06 GnPR - 0.015 HBDG + 1.311E - 05 WALL\left(m^2\right) + 0.633 SVF \quad (9.4)$$

$$T_{min}\left(°C\right) = 2.116 + 0.932\, Ref\, T_{min}\left(°C\right) - 0.272 GnPR + 0.003 PAVE - 0.004 HBDG + 1.619E - 05 WALL\left(m^2\right) - 1.33 ALB \quad (9.5)$$

Additional climate predictors in the enhanced models include maximum of daily wind speed ($WIND_{max}$), and daily average temperature at reference point for daytime ($T_{avg(daytime)-r}$) and nighttime ($T_{avg(nighttime)-r}$). An additional urban morphology predictor in the enhanced models is average height of buildings (AvgHeight).

In its initial development, the STEVE tool focuses on predicting outdoor temperature due to surrounding conditions. Afterwards, the numerical results are transformed into a graphical presentation in the form of a temperature map; which is deemed to be more presentable and informative for non-scientist users (e.g. planners or architects).

2.1 Outdoor Thermal Comfort Prediction Model

Thermal Sensation Vote (TSV) is an outdoor thermal comfort prediction model proposed for Singapore; which was developed based on various statistical techniques (Yang et al. 2013a, b). The proposed TSV model is applicable to outdoor urban spaces in Singapore and is different from the Predicted Mean Vote-Predicted Percentage Dissatisfied (PMV-PPD) model developed by Fanger (1973) which is prescribed for indoor spaces.

The TSV model was based on a very large scale survey from more than 2000 occupants conducted in 13 different outdoor urban spaces such as parks, gardens, river side, squares, university campuses, etc. Air temperature, relative humidity, wind speed, globe temperature and solar radiation were measured at the surveyed places.

It was found that outdoor thermal comfort can be improved by means of appropriate urban design since street orientation, aspect ratio (H/W) and vegetation were

Table 9.3 TSV index categories of outdoor thermal comfort

TSV range	Perception
2 ~ 3	Warm to hot
1 ~ 2	Slightly warm to warm
0 ~ 1	Neutral to slightly warm
−1 ~ 0	Slightly cool to neutral
−2 ~ −1	Cool to slightly cool
−3 ~ −2	Cold to cool

all found to affect human thermal comfort in outdoor urban spaces. Hence, this model can be applied by urban designers to evaluate the thermal sensation of users under certain outdoor thermal environment (Ignatius et al. 2015).

For the plugin, the initial model was adapted and customized; mainly because it might not be feasible for the planners to obtain data such as mean radiant temperature. Hence, the adapted model uses air temperature (T_a) and wind speed (V) as the independent variables. The model can be written as:

$$TSV = 0.315T_a\left({}^{\circ}C\right) - 0.078V\left(m/s\right) - 8.825 \tag{9.6}$$

The TSV calculation output represents an index that can be categorized based on different human perceptions of comfort, which can be seen in Table 9.3, where the values range from −3 to 3, representing different human perceptions from cold to hot.

Considering the large sample size, different thermal comfort survey locations, and the high R squared for the regression model, the TSV model can be applied for other outdoor urban spaces in Singapore. A validation study was carried out to validate the TSV model which shows high coincidence between the modeled and the actual results. However, the TSV model may not be applicable to other cities or countries due to different climatic conditions and different thermal adaptabilities of people.

3 Mapping Tool: STEVE Plug-in for SketchUp

3.1 Developing Analysis Tool Plugin for SketchUp

The concept of developing an integrated analysis tool derives from the notion that despite the promising developments of research on microclimate and built environment, the main issue lies in the integration of the two aspects in the design process. This is because most designers perceive that responsibility for mitigating the environmental problems is not part of the design process. Hence, this results in not just a detachment between research and industry practice, but also in fragmented research in sustainable design (Wong 2015).

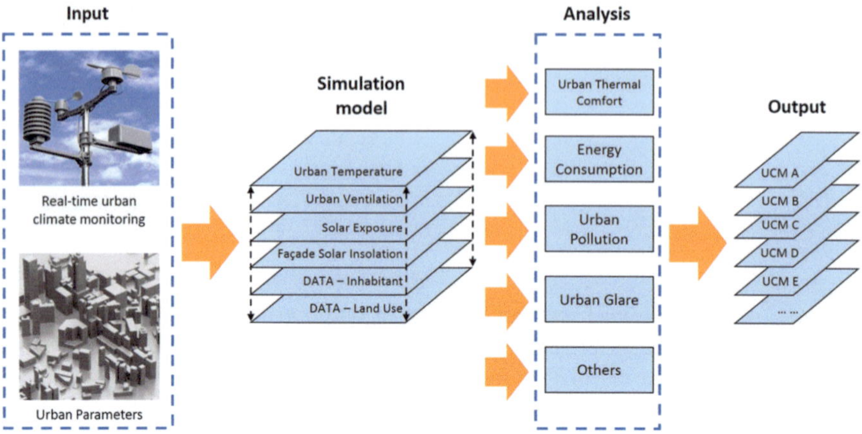

Fig. 9.3 Schematic on the integrated microclimate analysis tool

Figure 9.3 shows the workflow of integrating multifaceted relationship between microclimate and built environment in urban area. In its initial development, the prediction models focused on predicting outdoor conditions due to surrounding condition. The initial process of this type of analysis has always been looking at real-time urban climate monitoring and urban parameters. Afterwards, the numerical results were calculated and analyzed, then transformed into graphical presentation in form, using the UCmap approach. This was deemed to be more presentable and informative for non-scientists users (e.g. planners or architects).

At this stage, both THE STEVE and TSV prediction models have been embedded into a plugin for SketchUp, which has further improved the interoperability experience for planners, since the software has been well acknowledged as a common 3D modeler software in the industry. Hence, the required urban area model can be easily obtained as data input for models calculation.

3.2 Preparing and Exporting 3D Models from SketchUp

The conjunction between the STEVE tool and SketchUp further improves the interoperability experience for planners because SketchUp has been well acknowledged as a common three-dimensional (3D) modeler software in the industry. Hence, the required urban area model can be easily obtained as data input for STEVE calculation.

Figure 9.4 shows the workflow of the plug-in, which requires two inputs, i.e. 3D model from SketchUp and background climate. The 3D model from SketchUp requires input on buildings' shape (with enclosed top and bottom envelope), roads or pavement, and greenery (trees, shrubs or grass), as shown in Fig. 9.5. A 23-ha CBD area in Singapore has been selected as an example for the plugin to run the

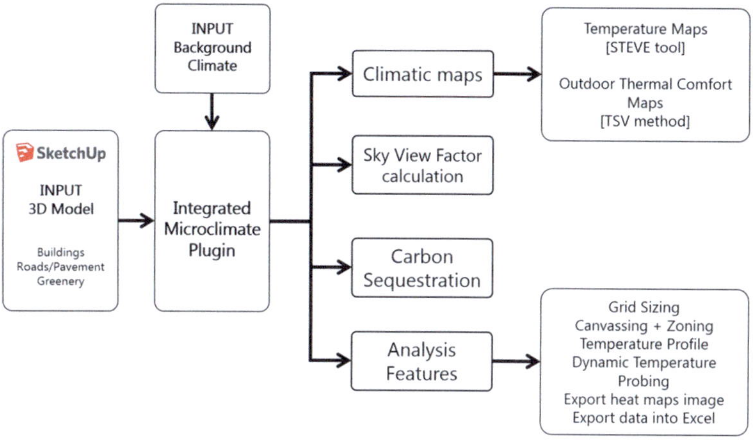

Fig. 9.4 Workflow of STEVE Tool plug-in

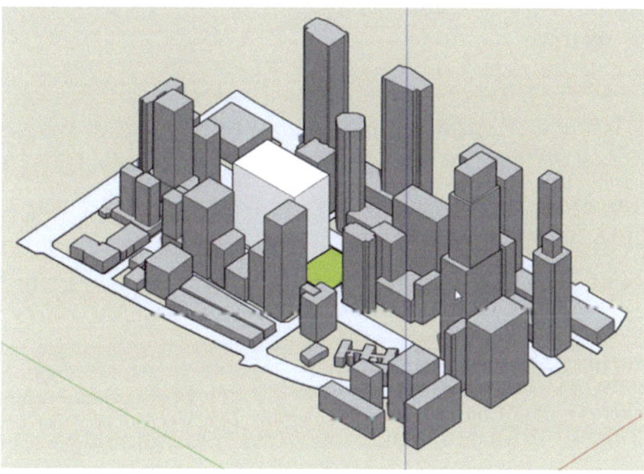

Background climate

Ref T_{min}	25.70	°C
Ref T_{avg}	28.08	°C
Ref $T_{avg\text{-}day}$	29.38	°C
Ref $T_{avg\text{-}night}$	26.78	°C
Ref T_{max}	31.50	°C
$SOLAR_{max}$	927.22	W/m^2
$SOLAR_{total}$	6,549.61	W/m^2
Albedo	0.16	
Wind speed	0.69	m/s

Fig. 9.5 (*left*) An example of 3D model within SketchUp comprises buildings, roads, and turfing. (*right*) Background climate condition for reference input

analysis. In addition the user can modify the background climate according to urban planning and design needs. These background climate parameters are mostly obtained from the weather station, which characterizes the microclimate condition in the mesoscale.

3.3 *Sky View Factor Calculation*

The Sky View Factor (SVF) is the percentage of a point's field of view that is occupied by the sky as opposed to buildings, trees or any other object in the landscape (Schneider 1992). It is a dimensionless parameter with values between zero and one. For instance, an urban canyon with a large aspect ratio will have a small SVF; while flat fields with an unobstructed horizon will have a value closer to one. The SVF plays a significant role in urban climatology (Oke 1987), especially for evaluating heat island phenomena in big cities (Unger 2009).

The SVF calculation feature within the plugin was developed with a calculation method using digital models of the environment (Matuschek and Matzarakis 2010; Matzarakis and Fröhlich 2010; Matzarakis et al. 2007). Calculation of SVF is based on knowledge of each angle element of the hemispheric environment and associated elevation (in connection with produced shadow) angle β and azimuth angle α. Therefore, the SVF can be assumed to be the sum of all this angle information over the whole hemispheric environment (see Eq. 9.7).

$$SVF = 1 - i\sum \sin^2 \beta_i \left(\frac{a_i}{360°} \right) \tag{9.7}$$

3.4 Outdoor Temperature Maps

The main products of the STEVE tool calculation are the temperature maps: T_{max},T_{avg}, T_{min}, $T_{avg\ (daytime)}$, and $T_{avg(nighttime)}$,. Users are also able to customize the cell sizing for the XY grid, which determines the resolution level of the produced maps. Figure 9.6 showcases the temperature maps using a 5-meter grid cell size, which adequately shows the hot and cool spots in both day and night time in a selected urban area.

The user can also utilize the temperature profile analysis feature which shows a sectional profile showing all temperature conditions (see. Fig. 9.7). Hence, one can observe the temperature behaviour in different urban canyons. No calculation is available for a grid cell that falls entirely inside a building, hence it is indicated with gaps within the sectional graph. Using the sectional profile, any hot spots along the sectional line can be identified. A function to locate the coordinate of the grid is available.

3.5 Greenery Planting Tool Feature

Information on different types of plants with their carbon storage and sequestration data based on a database from the National Parks Board (Tan and Sia 2009; Tan et al. 2009) has also been added into the plug-in. Through SketchUp, different plants can be added into the site planning; the carbon sequestration is then calculated accordingly.

An example of greenery planting within the 3D model in SketchUp is shown in Fig. 9.8, where users can pick the type of tree based on the various categories, spe-

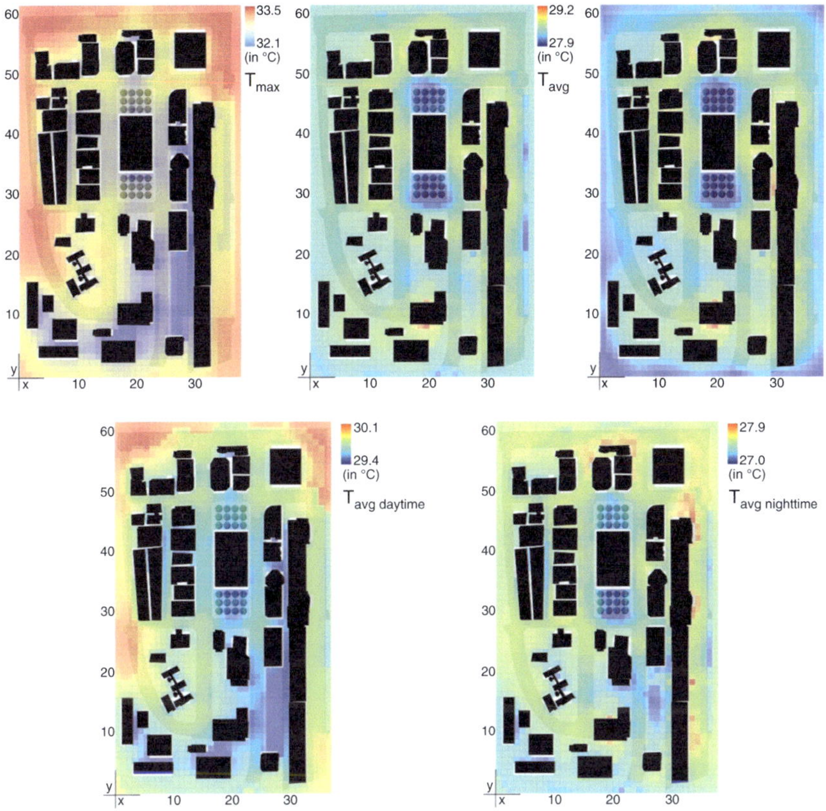

Fig. 9.6 Various temperature maps

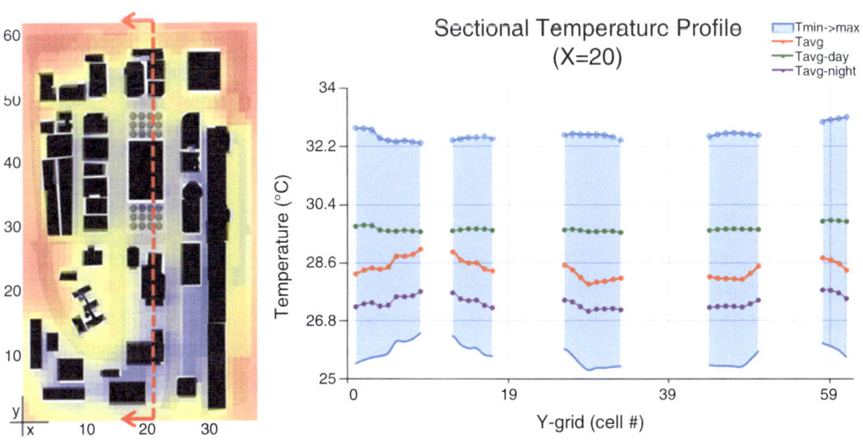

Fig. 9.7 Sectional temperature profile

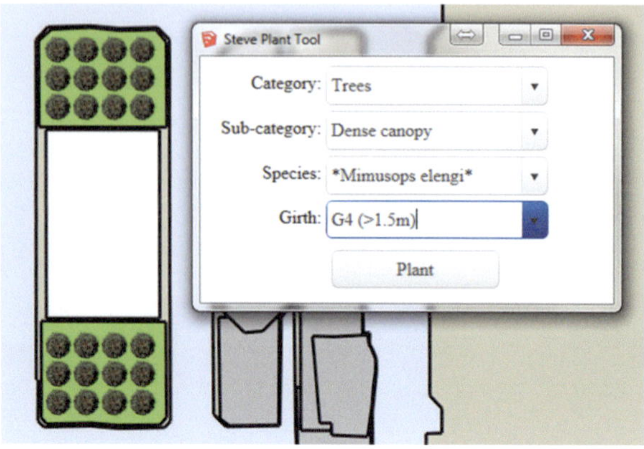

Fig. 9.8 Planting of point based greenery such trees and turfing in SketchUp

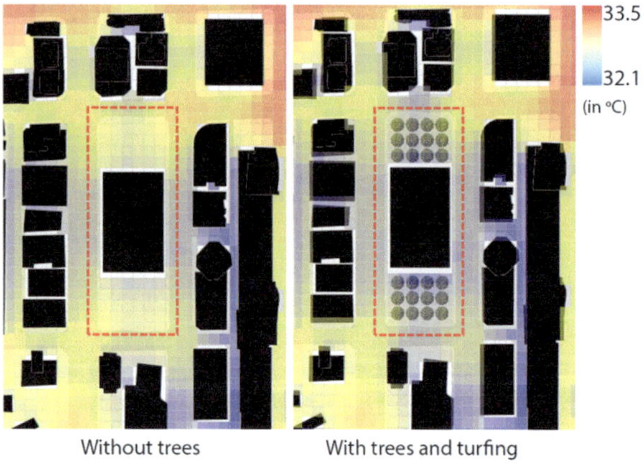

Fig. 9.9 Temperature T*max* maps comparing master plan with and without trees

cies, and girth size. Trees are denoted with points, while grass and shrubs are ground covers, which use polygonal shapes.

Figure 9.9 shows temperature maps comparing the effect of greenery on the temperature in an urban area. One can observe the reduced maximum temperature (T_{max}) due to the presence of dense canopy trees in the building's courtyard. Users can select each of the grid cells to obtain information on the green plot ratio (GnPR) for both trees and grass cover. Furthermore, each plant has its own carbon information, which is used to calculate the overall carbon storage and sequestration of a selected area (see Fig. 9.10).

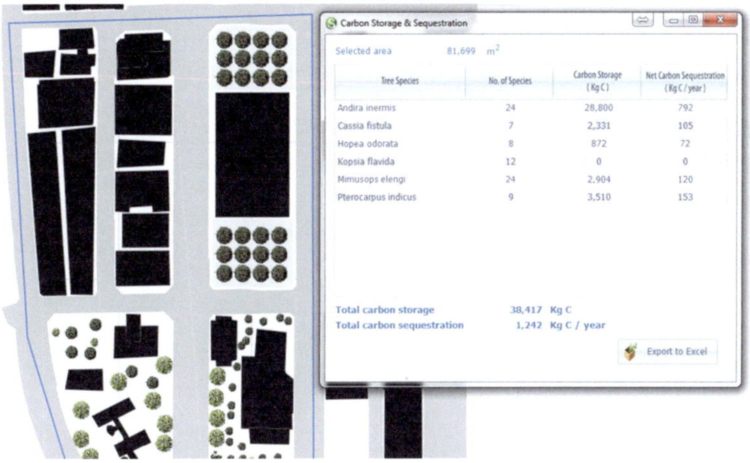

Fig. 9.10 Carbon storage and sequestration of the selected area

3.6 Outdoor Thermal Comfort Maps

While T_a is obtained from the STEVE tool calculation e.g. T_{max}, T_{min}, T_{avg}, etc., the wind speed value can be customized for the whole area automatically, or defined by inputting a different value for each grid cell. One can perform a comparison study of different wind conditions to see the impact on the outdoor thermal comfort; particularly in urban canyons or open spaces. Figure 9.11 shows different TSV maps in parallel with different wind speed values from the same temperature output (in this case looking at the peak day time condition or T_{max}).

Referring to Table 9.3, the increase in wind speed within highly dense urban areas could provide better comfort for the pedestrians especially in the open spaces. Overall, one can observe that the presence of wind speed of 2 m/s at the hottest period of the day (T_{max}) could lower the average TSV perception from 1.6 to 1.2 compared to stagnant wind condition.

It is important to note that the prediction model calculates the impact of wind speed empirically toward the temperature changes (not based on computational fluid dynamics concept), and it does not consider the wind direction aspect.

Figure 9.12 compares different scenarios regarding trees, iterating from none to dense planting. The comparison shows how the plugin is able to simulate the effect of trees on outdoor temperature, and consecutively the outdoor thermal comfort due to the shading effect. The figure illustrates the TSV maps under the T_{max} condition and similar wind speed condition. Furthermore, open spaces not occupied by buildings or roads were considered as grass surface (leaf area index or LAI = 1). Table 9.4 displays the comparison of different GnPR conditions resulting from tree planting iteration, where it shows the gradual decrease of the TSV during T_{max} condition due to shading from the trees.

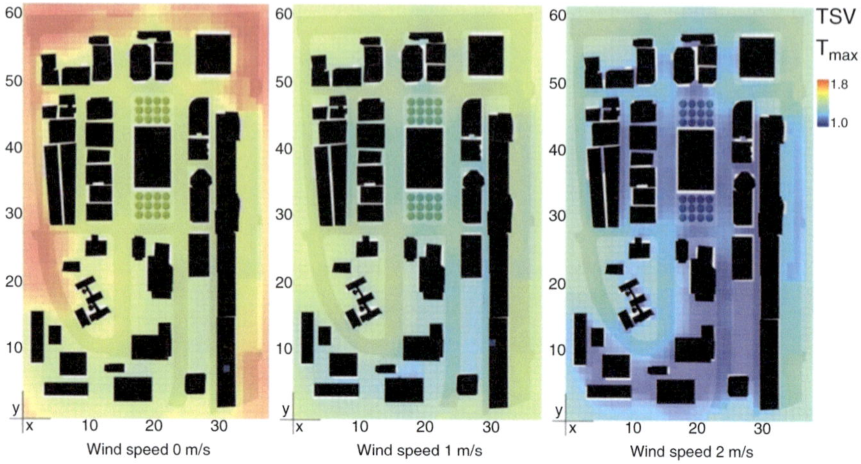

Fig. 9.11 TSV maps for T_{max} comparison, with wind speed iteration from 0–2 m/s

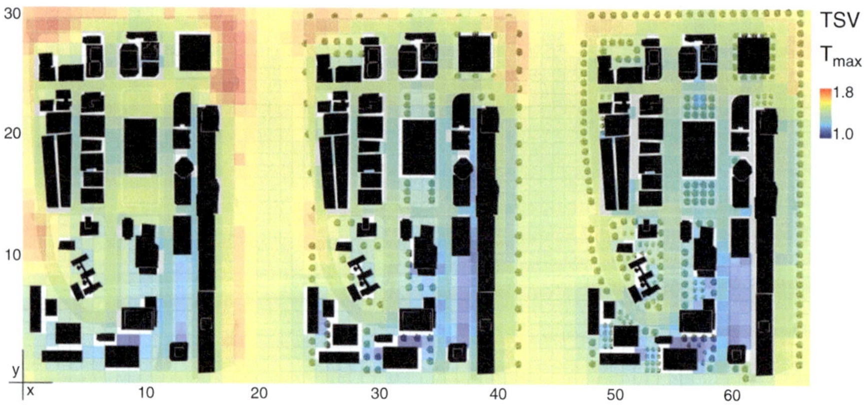

Fig. 9.12 TSV maps for Tmax comparison, with tree planting iterations, from none to dense

Table 9.4 TSV comparison from different GnPR condition due to trees planting

Scenario	1	2	3
Description	No trees	Decent trees	Dense Trees
GnPR	0.83	1.21	1.43
Maximum TSV T_{max}	1.69	1.64	1.62
Average TSV T_{max}	1.51	1.48	1.46
Minimum TSV T_{max}	1.25	1.22	1.20

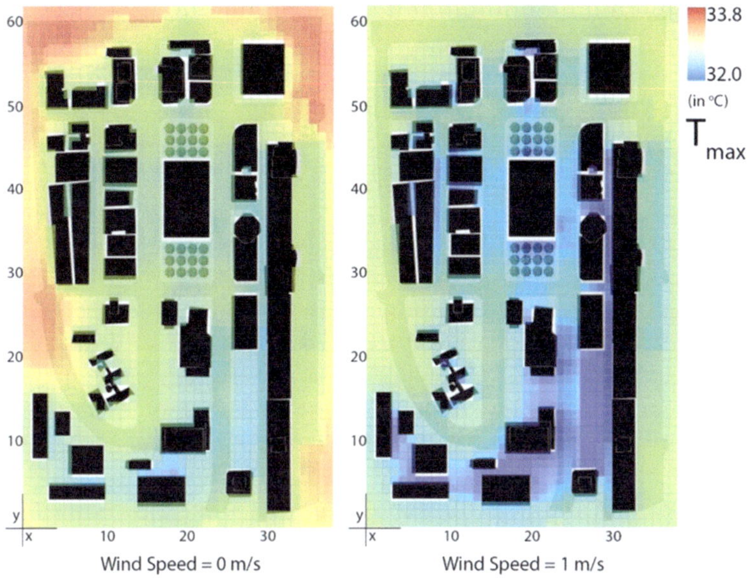

Fig. 9.13 Temperature T*max* maps comparing master plan with different wind speeds

3.7 Wind Speed Consideration

The current STEVE tool model has also considered the impact of wind speed in predicting the ambient temperature, particularly for the T_{max} output. Within the plugin, users are able to customize the wind speed value for the whole area automatically, or define a different value for each grid cell.

Figure 9.13 illustrates the comparison of an urban area with and without considering the wind speed aspect. It is important to note that the prediction model calculates the impact of wind speed empirically on the temperature changes (not based on computational fluid dynamics concept), and it does not consider the wind direction aspect.

3.8 Data Exporting

For data output, users can obtain data in two different formats: graphic/image file (.JPG format) and data tabulation (.CSV format) for further analysis. Hence, various urban environmental solutions can be examined by the researchers or urban planners directly through graphical representation (maps) or through data (numbers).

4 Conclusion

In order to carefully preserve the quality of urban environment, temperature mapping method can be utilized to examine the effectiveness of various urban environmental solutions in planning for future city developments. STEVE Tool was developed with a motivation to link research findings, especially air temperature prediction models, with urban planners. It is further integrated with SketchUp to serve as a user-friendly platform for urban planners to be well informed regarding the master plan and its impact on the microclimatic condition.

Such information is deemed necessary to ensure that the master plan and its surroundings do not suffer additional temperature increase which can worsen the urban heat island phenomena. This mitigation approach at the early stage of the design process can be proven beneficial especially to tackle the environmental issues earlier rather than at later stages. Furthermore, due to the minimal time required to produce such maps, benchmarking between different design options can be done simultaneously, hence different stakeholders can discuss and analyze multiple master plan options at the same time.

5 Future Work

This article presents the capability of the integrated analysis tool for looking at outdoor temperature and thermal comfort performance of an urban area. For the future work, the thermal load aspect will be considered as part of the plugin, to simulate the external heat gains and sensible cooling load due to urban layout. The application of these different microclimate aspects has been shown in a previous study, which utilized the STEVE tool, TSV, and thermal load components in analyzing different master plan solutions (Ignatius et al. 2015).

Acknowledgement The study is part of the research project "Development of Climatic Mapping Tool for Estate Environmental Evaluation" funded by Jurong Town Corporation (JTC) Singapore under NUS – JTC Industrial Infrastructure Innovation Centre. The STEVE Tool plugin for SketchUp was developed by nForce Solutions Pte. Ltd.

References

Alfaro EJ, Gershunov A, Cayan D (2006) Prediction of summer maximum and minimum temperature over the Central and Western United States: the roles of soil moisture and sea surface temperature. J Clim 19(8):1407–1421

Bottyán Z, Unger J (2003) A multiple linear statistical model for estimating the mean maximum urban heat island. Theor Appl Climatol 75:233–243

Fanger PO (1973) Thermal comfort: analysis and applications in environmental engineering. McGraw-Hill, New York

Giridharana R, Laua SSY, Ganesana S, Givonib B (2007) Urban design factors influencing heat island intensity in high-rise high-density environments of Hong Kong. Build Environ 42:3669–3684

Ignatius M, Wong NH, Jusuf SK (2015) Urban microclimate analysis with consideration of local ambient temperature, external heat gain, urban ventilation, and outdoor thermal comfort in the tropics. Sustain Cities Soc 19:121–135

Jusuf SK, Wong NH (2009) Development of empirical models for an estate level air temperature prediction in Singapore. In: Second international conference on countermeasures to urban heat Islands, Berkeley, United States

Jusuf SK, Wong NH (2016) Valuing green spaces as a heat mitigation technique. In: Santamouris M, Kolokotsa D (eds) Urban climate mitigation techniques. Routledge, London, pp 41–62

Katzschner L (1988) The urban climate as a parameter for urban development. Energ Buildings 11:137–147

Katzschner L, Mülder J (2008) Regional climatic mapping as a tool for sustainable development. J Environ Manag 87:262–267

Kolokotroni M, Giannitsaris I, Watkins R (2006) The effect of the London urban heat island on building summer cooling demand and night ventilation strategies. Sol Energy 80(4):383–392

Lo CP, Quattrochi DA (2003) Land-use and land-cover change, urban heat island phenomenon, and health implications: a remote sensing approach. Photogramm Eng Remote Sens 69(9):1053–1063

Matuschek O, Matzarakis A (2010) Estimation of sky view factor in complex environment as a tool for applied climatological studies. In: Matzarakis A, Mayer H, Chmielewski F (eds) 7th conference on biometeorology. Meteorological Institute, Albert-Ludwigs-University of Freiburg, Freiburg, pp 535–540

Matzarakis A (2005) Country report: urban climate research in Germany, pp 4–6

Matzarakis A, Fröhlich D (2010) Skyhelios. Meteorological Institute, Albert-Ludwigs-University Freiburg, Freiburg

Matzarakis A, Rutz F, Mayer H (2007) Modelling radiation fluxes in simple and complex environments – Application of the RayMan model. Int J Biometeorol 51:323–334

Ng E, Chao R, Katzschner L, Yau R (2009) Urban climatic studies for hot and humid tropical coastal city of Hong Kong. In: The 7th international conference on urban climate, Yokohama

Oke TR (1987) Boundary layer climates. Routledge, London

Ren C, Ng E, Katzschner L (2011) Urban climatic map studies: a review. Int J Climatol 31:2213–2233

Rohington E (2005) Thermal comfort implications of urbanization in a warm-humid city: the Colombo Metropolitan Region (CMR), Sri Lanka. Build Environ 40(12):1591–1601

Schneider SH (1992) Introduction to climate modeling. In: Climate system modeling. Cambridge University Press, Cambridge, MA, pp 3–26

Taha H, Chang S-c, Akbari H (2000) Meteorological and air quality impacts of heat island mitigation measures in three U.S. cities. Lawrence Berkeley National Laboratory, Berkeley

Tan PY, Sia A (2009) Leaf area index of tropical plants. National Parks Board, Singapore

Tan PY, Yeo B, Yip WX, Lua HS (2009) Carbon storage & sequestration by urban trees in Singapore, centre for urban greenery and ecology. National Parks Board, Singapore

Tanaka T, Moriyama M (2011) Urban environmental climate maps for supporting environmental and urban planning works in local government. AIJ J Technol Des 17(36):687–692

Unger J (2009) Connection between urban heat island and sky view factor approximated by a software tool on a 3D urban database. Int J Environ Pollut 36:59–80

Wong NH (2015) Grand challenges in sustainable design and construction. Frontiers Built Environ 1

Wong NH, Jusuf SK (2008) GIS-based greenery evaluation on campus master plan. Landsc Urban Plan 84:166–182

Wong NH, Jusuf SK, La Win AA, Thu HK, Negara TS, Wu X (2007) Environmental study of the impact of greenery in an institutional campus in the tropics. Build Environ 42:2949–2970

Wong NH, Jusuf SK, Syafii NI, Chen Y, Hajadi N, Sathyanarayanan H, Manickavasagam YV (2011) Evaluation of the impact of the surrounding urban morphology on building energy consumption. Sol Energy 85:57–71

Yang W, Wong NH, Jusuf SK (2013a) Thermal comfort in outdoor urban spaces in Singapore. Build Environ 59:426–435

Yang W, Wong NH, Zhang G (2013b) A comparative analysis of human thermal conditions in outdoor urban spaces in the summer season in Singapore and Changsha, China. Int J Biometeorol 57:895–907

Chapter 10
Sustainable Fishing Settlement in Muara Angke, North Jakarta

Tri Harso Karyono, Dedi Burhanudin, and Benedict Timothi

Abstract Global warming is melting the polar ice, raising the sea levels. Frequent floods occur in many coastal cities, including coastal settlements, even on sunny days. The Muara Angke fishing settlement in North Jakarta is no exception. Flooding is occurring more frequently at any time, resulting serious problems for the entire inhabitants. The floods come from the sea water during high tides and also from the overflow of the river water during the rainy season. A number of people have tried to escape from the flooding by raising the floors of their houses, the rest, who are very poor and have no money to do anything, just pray that nature may not send them more water. Within the last few years the Jakarta authority took action to evict them as they are occupying illegally lands belong to the government. This paper proposes a different way, not to evict the inhabitants, but rather to develop a new elevated fishing settlement in this area to avoid frequent floods. Sustaining the fishermen in this area is important as they have been collecting the fish in a traditional way for quite a long time, providing fish for the people.

Keywords Elevated settlement • Eviction • Fishing • Flood • Stilt houses • Sustainable design

1 Introduction

Due to global warming and climate change, settlements in coastal areas have suffered from increasingly frequent floods. A report from the Intergovernmental Panel on Climate Change (IPCC) in 2013 showed that within the period of 1901–2010, global mean sea level rose by 0.19 (0.17–0.21) m. The floods could take place at any time, even on sunny days, not only during heavy rain. A fishing settlement in Muara Angke, North Jakarta, has suffered a serious problem with floods occurring

T.H. Karyono (✉) • D. Burhanudin • B. Timothi
School of Architecture, Tanri Abeng University, Jakarta, Indonesia
e-mail: t_karyono@yahoo.com

T.H. Karyono et al. (eds.), *Sustainable Building and Built Environments to Mitigate Climate Change in the Tropics*, DOI 10.1007/978-3-319-49601-6_10

more frequently than before. A number of people have tried to escape from the floods by raising the floors of their houses, however, the rest of the inhabitants, who cannot afford to modify their houses, just pray for mercy, that nature will not send them too much water from the sea and the river. Currently, the local authority has yet to take any serious action to solve the problems regarding the flooding. On the contrary, the authority is even preparing an action to evict the inhabitants of the area. They will be relocated in multi-storey flats to be built by the authority in the future.

As happens in many developing countries, such as Indonesia, the gap between the poor and the rich is quite wide. The poor are becoming poorer because in many cases the government does not provide an easier access to opportunities to improve their life. Over a long period, most of the developing countries such as Indonesia have suffered from corruption, in which most of the state's revenues are lost to individuals, mainly government officers, either in direct or indirect ways. Because of this, the government tends to have a limited budget to improve the facilities required by the public, such as education, healthcare facility and housing, particularly for the poor. In the developing countries, it seems that the rich only do less to get more, while the poor have to do more to get less. The Muara Angke people tend to work very hard from dawn to night, mostly everyday, but this is only enough to feed them with basic meals with nothing left at the end.

In terms of the housing sector, the fishing settlement in Muara Angke is only a small part of the housing problems that this country has. Although housing problems take place in any part of the world, regardless of the country's economic level, Indonesian housing problems are likely to deal with the inadequate number of houses available to shelter the people in need. Poor people are unlikely to have a chance to buy or rent proper houses. The homeless tend to stay in their relatives' houses if available, otherwise, they occupy the slum areas. One problem of living in a slum area, such as in Muara Angke, is that the residents have to prepare for being evicted at any time (Karyono 2004).

2 Muara Angke Fishing Settlements

Muara Angke is a popular fishing venue for the Jakarta people. This fishing area is located at the seaside on the Java Sea in the North Jakarta municipality. A fishing settlement has occupied a part of this area for a long time. There is a fish market, not far from this settlement. In the Indonesian language, the word 'muara' literally means a river-mouth. Since this area is located around the river-mouth of Angke River, it has become known popularly as Muara Angke (Wikipedia 2016).

Figure 10.1 shows the area of Muara Angke and its squatter settlement, inhabited mostly by fishermen. Close to the squatter settlement, there is Muara Angke port, which provides public transport by boat to the thousand islands, about 15–60 min away from this port. Close to the port, there are a big fish market and some fish restaurants, which are popular with the Jakarta people.

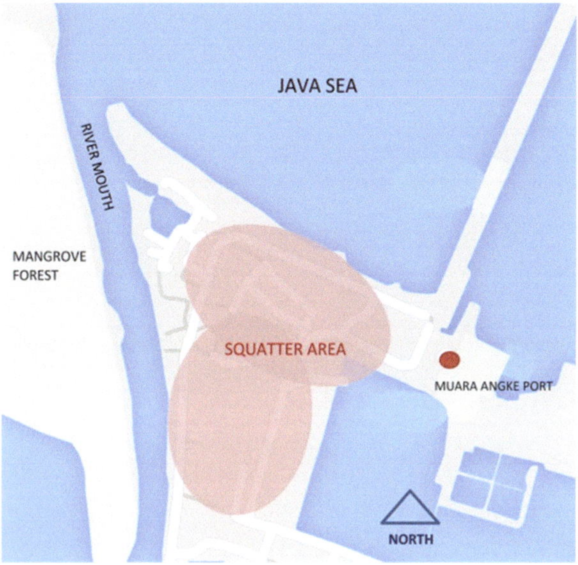

Fig. 10.1 Muara Angke squatter settlement

Muara Angke is surrounded by a river on the west side and the Java Sea in the north. Across the river to the west, there is a mangrove forest with a size of around 25 ha, inhabited by many bird species. This forest is a part of the mangrove forest remaining in the Jakarta region (Garuda 2016). Indonesia has the biggest mangrove forest area in the world. A few years ago the total size of the Indonesian mangrove forest was 4.25 million ha or about 27 % of the total world forest (Alikodra 1996). However, this size has been reduced gradually due to a number of coastal developments in many parts of Indonesia.

Regarding the local climate, this area has relatively high daily air temperature. The climatic variations in this area can be referred to the closest weather station located at the Tanjung Priok Harbor, about 15 Km away. The twenty four hours recording in 13 January 2016 reported that the range of air temperature was between 27 and 33 °C, relative humidity was between 55 and 79 % and the wind speed was between 3 and 15 knots (5.6 and 27.8 km/h), mainly from northwest and north direction (Jakarta Climate Office 2016).

Although Muara Angke was originally inhabited by a fishing settlement, this area is now occupied by a number of people with different occupations other than fishing. Along with the urbanization process, Muara Angke is now becoming a slum area. The inhabitants have built shelters with light and non permanent materials, such as cartons, wooden boards, steel sheets and plastics. There are a number of problems facing the inhabitants. Not only are they lacking proper sanitation and clean water, but also they experience frequent floods which can occur at any time, regardless of the seasons. The floods are caused by the high tides of the sea water and the overflow of the river water nearby. Other than floods, the inhabitants also are

Fig. 10.2 Fishing boats moored at Muara Angke beach with a background of modern apartments built on the reclamation land

threatened with eviction as they have occupied the government land. Figure 10.2 shows a number of fishing boats moored in Muara Angke beach with a background of modern apartments, built by private developers on the reclamation land. The current fishing settlement is likely to be evicted and to be extended by land reclamation. As happened in the previous reclamation around this area, the extended land is likely to be used for the same purpose as before: modern facilities for the rich, such as exclusive apartments, luxurious houses, shopping malls, etc.

3 The Inhabitants' Profiles and the Actions Taken to Reduce the Floods

Due to the rise of sea water and high tides, the Muara Angke settlement has suffered from frequent floods, which occur at any time regardless of the season. To have indepth information about the profile and the problems facing the inhabitants in this squatter settlement, we conducted a field survey in March 2015. Two further site visits were made in 2016 to see the eviction progress and to take some more pictures. Only a proportion of the inhabitants were available to be interviewed, however, the small data colllection is likely to reflect the majority of people living in this settlement. Thirty one households were visited for interviewing.The inhabitants comprised 11 males and 20 females, aged between 21 and 67 years. Table 10.1 shows

Table 10.1 Respondents' occupations

Occupation	Number of respondents
Fisherman (male)	15
Fish supplier	1
Street vendor	5
Daily wage labourer	5
Small company worker	2
Scavenger	1
Teacher	1

Table 10.2 Respondents' family members

No of family members	Number of respondents
1	1
2	6
3	9
4	8
5	2
6	1
7	1
8	1
11	2

Table 10.3 Respondents' length of stay in the current houses

1	5
2–5	6
6–10	6
11–20	12
30	2

the respondents' occupations, recorded from the survey. Out of 30 respondents, 15 of them were fishermen, one was a fish supplier, five were street vendors, selling dairy products and cooked foods, five were daily labourers, two were working in companies, there was one scavenger and one teacher.

Table 10.2 shows the number of family members of the respondents. The range of the members was between 1 and 11, however, the three most common numbers of the family members were 2, 3 and 4 persons, with the number of respondents being 6, 9 and 8 respectively.

Table 10.3 shows the length of stay of respondents in their houses, five respondents had been staying for 1 year, six respondents staying between 2 and 5 years, another 6 respondents had been in their houses between 6 and 10 years, with 12 respondents staying between 11 and 20 years, while two respondents had been in their houses for 30 years.

Table 10.4 Respondents' house sizes

Size of the house (m^2)	Number of respondents
12	7
15	1
18	2
20	1
24	16
25	1
32	3

Fig. 10.3 An interior of one of the shelters built withlight and non-permanent materials

Table 10.5 Average number of floods entering the house experienced by the respondents annually

Number of floods	Number of respondents
1	8
2	24
3	2
n/a	9

Table 10.4 shows the size of respondents' houses. There were seven respondents who lived in houses of 12 m^2, one respondent lived in a house of 15 m^2, two respondents stayed in houses of 18 m^2, one respondent lived in a house of 20 m^2, 16 respondents occupied houses of 24 m^2, one respondent lived in a house of 25 m^2 and three respondents occupied houses of 32 m^2. Figure 10.3 shows an interior of one of the shelters built withlight and non-permanent materials.

Table 10.5 shows the average number of floods that the respondents had experienced annually. There were eight respondents who had experienced having one flood a year, 24 respondents had flooding twice a year and two respondents had three a year, while nine respondents could not remember how many floods they have annually.

Table 10.6 Number of times respondents raising the floors throughout the time they lived in this settlement

No of times floors are raised	Number of respondents
1	11
2	2
3	4
4	6
5	1
n/a	7

Table 10.7 The height of floors being raised

Height of the raised floor (cm)	Number of respondents
15	2
30	5
35	3
40	9
45	2
60	3
n/a	7

Table 10.6 shows the number of respondents who had been raising the floors of their houses throughout the time they lived in this settlement. This action was taken in order to minimise the impact of the floods. There were 11 respondents who have raised their floor once, two respondents have done it twice, four respondents made it three times, six respondents have done it four times and one respondent has done it five times. Seven respondents provided no data about this matter.

Table 10.7 shows the height that the inhabitants have raised their floors in order to minimize the impact of the floods. Two respondents raised the floor up to 15 cm, five respondents raised up to 30 cm, three respondents have done up to 35 cm. Nine respondents have raised the floors up to 40 cm, two raised up to 45 cm and three respondents raised the floor up to 60 cm. Seven respondents were not available to present the data regarding this matter.

In terms of the dwelling design, five out the the dwellings were designed as two storey houses and the rest were single storey. Ten dwellings have no bathroom and toilet, so the inhabitants use public toilets and bathrooms available in this area. Figure 10.4 shows the current shelters which are still existing and waiting for the demolition.

In terms of monthly income, Table 10.8 shows the income distribution of the respondents. The respondents' monthly incomes ranged between 500 thousand and 5 million rupiahs (US$38–US$385).

Since the Jakarta monthly minimum wage today is 3, 1 million rupiahs (US$ 235), a monthly income of less than 3 million would be considered as a low income. However, the lowest income of 500,000 rupiahs (US$ 38) is still adequate to feed

Fig. 10.4 Existing shelters, waiting to be demolished

Table 10.8 Monthly income of the respondents

Monthly income		
Indonesian rupiah	US dollar	Number of respondents
500,000 to <1,500,000	38 to <114	9
1,500,000 to 3,000,000	114 to 228	14
3,000,000 to 5,000,000	228 to 385	8

two persons, including water and basic electricity if they do not pay for the shelter or the transport. An income of between 3 and 5 million, in this case, is considered to be a moderate income.

4 Problems of the Settlement and Inhabitants: Floods and Evictions

Frequent floods from the sea water take place at any time when there is a high tide. With poor infrastructure, the water remains in this area for days and even weeks, creating problems with transportation and creating an unhealthy environment. Since this area is a squatter settlement, in which the land belongs to the Jakarta authority, no action has been taken so far by the authority to improve it. Figure 10.5 shows the current condition of the access road to the settlement, in which the road is always inundated by the sea and rain water.

Fig. 10.5 An access road to Muara Angke settlement is always inundated by the sea and rain water

From the interviews, twenty respondents admitted that they possessed the shelters where they lived, while 11 respondents said they rented the shelters from the owners. However, all the inhabitants admitted that their shelters have occupied government land. For this reason, the Jakarta authority has warned the inhabitants that they will evict them.

The authority is still considering to build a number of multi-storey flats to accommodate them, however, the inhabitants tend to reject the government plan to remove them. The reasons for the rejectionis that they want to sustain their current jobs. Also, they have already adapted to the current environment and their neighbours, so that it is hard for them to leave their shelters, eventhough they have to face a problem of frequent floods all the time. They prefer to stay in their current settlement whatever the conditions. For them, if there is some assistance offered by the government or private sector, their preference is firstly, they want their houses to be saved from the floods, secondly, they want to have better sanitation for the settlement and they want to have a clean water facility in the settlement.

There has been a plan by the private sector in collaboration with the Jakarta authority to expand this area by means of land reclamation. Under this plan this settlement must be evicted and demolished. The reclamation land will be used mainly for commercial purposes, such as modern apartments, shopping malls and luxurious houses.

As compensation stated in the government regulations, the private company has to provide a small portion of the new reclaimed land for the evicted inhabitants currently living in the Muara Angke settlement. The company promises to build a number of five-storey flats to accommodate the evicted families of Muara Angke. However, this plan has been abandoned since there are a number of rejections from the environmentalists and non government organisations for land reclamations. The reason is that reclamation has destroyed the marine biota, reducing the number of fish in the nearby sea and creating problems for the fishermen to collect fish.

Many land reclamations have been done before in the Jakarta bay by a mutual collaboration between the Jakarta authority and the private sector. Unfortunately, people have seen this as a hidden way to remove the poor for the advantage of the rich to provide a better place for them in the beautiful coastal areas. When the new reclamation plan was abandoned, the private company and the Jakarta authority declared they would cancel the five-storey flats for the Muara Angke inhabitants (Housing-estate.com 2016). However, the Jakarta authority has promised to build some multi-storey flats to accommodate the evicted inhabitants of Muara Angke (Jakarta Raya 2016). The problem is that these flats would be far away from the sea, which means the inhabitants would have to abandon their current profession as fishermen if they accept the authority's offers.

Just a few days before the end of Ramadan in mid-2016, the Jakarta authority has started to evict part of this settlement. Figures 10.6 and 10.7 show the recent eviction of the Muara Angke fishing settlement.The inhabitants were just giving up as a result of the authority's action. They did not have any idea where to go.

Many people suspect that this action is only a step forward to facilitating the rich, who live nearby this settlement, in the developed reclamation land. They want better access and views to the beach. Figure 10.8 shows two kids with an insecure future, playing around the shelters where they live.

Fig. 10.6 A view of the eviction process started a few days before the end of Ramadan in 2016

Fig. 10.7 Squatter shelters being demolished to allow the rich to have better access and view to the beach

5 Overcoming the Problems

Within the last decades, the Muara Angke settlement has suffered from frequent floods, which are likely due to the rise of the sea level and the overflow of the river water. Based on the field survey, a number of efforts have been made by the inhabitants to reduce the effect of floods, such as raising the floors of their houses. However, during the high tides the ground level of this area is about the same as the sea level. Therefore, there had been no chance for them to avoid the sea water entering their houses. Along with the floods, this settlement also has problems with poor sanitation and lack of clean water. To answer and to overcome these problems, we propose a plan to develop a new elevated settlement in the same area, by optimising an area of empty land which still exist the tip of this peninsula. Part of this empty land will be developed with a number of stilt houses. After they are completed, some of the inhabitants of the old settlement will be moving into these new houses. The old emptied shelters will be demolished and the next new houses will be built on their sites. Again, after completion some inhabitants in the old shelters would be moving into these new houses. This process is to be repeated again and again until all the inhabitants are accommodated in the new houses. More than 800 families are to be accommodated in the new settlement. We expect that the Indonesian government and also non-profit funded organisations will give their hands to support this project.

Fig. 10.8 Small kids playing around with insecure future, no playground available in this area

5.1 Flood Responsive Design

Considering that some areas of this settlement belong to the Jakarta authority, not to the inhabitants, there could be a possibility that the authority will take action to evict those who are considered to occupy land illegally. In order to sustain the living activities of all the families in this settlement, careful consideration needs to be taken rather than moving them away to the new flats, away from their current shelters. As happened before, this has created other problems for the evicted families, such as losing their current jobs, moving the kids away from the current schools, etc. In many cases, the authority tends to neglect the fact that the inhabitants are poor with a limited income to pay for the new rental flats. Also, there is a strong relationship amongst the community members in the settlement which is not easy to break.

Considering the problems of the poor people live in this area, we propose a plan to keep them staying in this area during any necessary improvement in their built environments. In order to keep them from the flood, a new settlement will be built with all of the houses and outdoor facilities on stilts. Clean water and appropriate sanitation will be provided so that the inhabitants will have a healthy environment. Furthermore, the surrounding area of the environment would be planted with mangrove and other trees to improve the air and sea water environments, as well as attract the wildlife. To electrify this settlement, a reasonable number of solar PV panels will be mounted around the settlement, including some on the roofs. In this study, we provide a very basic proposal of the development plan of the new settlement. Further details of the settlement facilities, building design and its materials need to be further investigated and analysed.

Learning from a number of traditional fishing settlements in Indonesia, such as in Sicini, Arungkeke Village, Jeneponto Regency, South Sulawesi (Patandianan and Toban 2011) and abroad (Fong 2014), all the facilities, including houses and outdoor areas of the new settlement of Muara Angke would be built on stilts. According to the IPCC, the global warming is expected to reach 3° at the end of this century, resulting in about a 1 m sea level rise (Girardet and Mendonca 2009). The maximum floor raising that the inhabitants have done in the past was 60 cm (Table 10.7). Considering the IPCC predictions and the highest elevation that the inhabitants have done, the new fishing settlement of Muara Angke will be elevated about 1.2 m higher than the current high tides which will make it higher than the water level of overflowing river water during the floods. The upper structure of the stilt buildings is supported by columns or poles made from concrete with circular section. The columns will support the roof structures made of wood.

The reason for using concrete poles instead of timber to support the stilt structures of the settlement is that Indonesia does not have an adequate timber product at the moment. The government tends to protect the forests so that good quality wood becomes expensive compared to concrete. Considering the durability of concrete compared to wood, it is likely that concrete would better resist the waves and would last longer than wood for this purpose. On the other hand, wood and bamboo are to be used for the superstructure elements, such as walls, doors, windows, roof structures and the floors of the houses. The open space decks and pedestrian walkways are also made from wood. The roofs of the houses are covered with local tiles and some parts are to be installed with solar PVs. Rain gutters are provided at all of the roofedges.

Water collected by the rain gutters is collected via a downspout into a water tank. All collected rainwater will be treated and used for household water, except for drinking and cooking. Both of these will still be supplied by the Jakarta water company.

5.2 *The Development Process of the New Fishing Settlement*

The proposed new fishing settlement is to be built in the area of the old settlement, by using a step-by-step redevelopment process. Only a small part of the current settlement will be redeveloped, serving as a primary model to be adopted by other fishing settlements in many parts of Indonesia. The size of the new settlement is about 10 ha, occupying the tip of the Muara Angke peninsula. On the top side of the peninsula, there is an area of open land which has not been occupied. This area will be used as the starting point for constructing the redevelopment of the settlement.

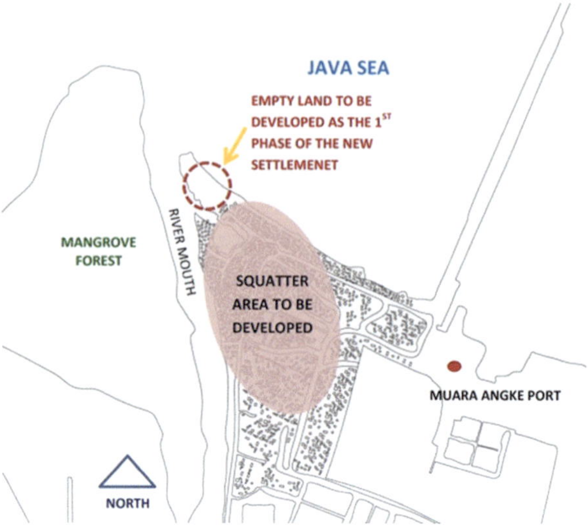

Fig. 10.9 Squatter area of Muara Angke and an area of empty land at the tip of the peninsula to be occupied for the first phase of the new settlement

The first step in the development is to build a number of new houses located on the open land. When the new houses are completed, a number of families in the old settlement will move into the new houses. The next step is to demolish all the vacated old shelters and prepare this now empty land for the next step of building the new houses. Again, the same process is repeated until all the inhabitants have occupied the new houses. The development steps of the new fishing settlement are shown in Figs. 10.9, 10.10 and 10.11.

The size of the shelters that the inhabitants use currently (Table 10.4) is between 12 and 32 m^2, while the majority are 24 m^2. On this basis the new dwellings are approximately adjusted to the current shelters' sizes. Those currently occupying shelters with the size of 12 and 15 m^2 would be provided dwellings with a size of 11 m^2 each. Those who occupy shelters between 18 and 25 m^2 would be provided with dwellings with a size of 21 m^2, while those with 32 m^2 and bigger would occupy dwellings with a size of 37 m^2 each. The proposed development comprises two types of buildings, Type A and B, which have exactly the same form, but with a different floor plan. The Type-A building consists of seven units of dwellings, in which the floor size of each unit is 37 m^2 (Fig. 10.12a). Each unit consists of two sub-units, with the floor area of 11 and 21 m^2, plus a public bathroom and toilet of 5 m^2 (Fig. 10.12b).

The Type-B building consists also of seven units of dwellings, in which the floor area of each unit is 37 m2, the same as Type-A, but it has no sub-unit. The entire floor of this unit is for a single family (Fig. 10.13).

In the centre of the building, there is an open space to be used for community common purposes, such as clothes drying, fish drying, kids' playground and a night time outdoor sitting area.

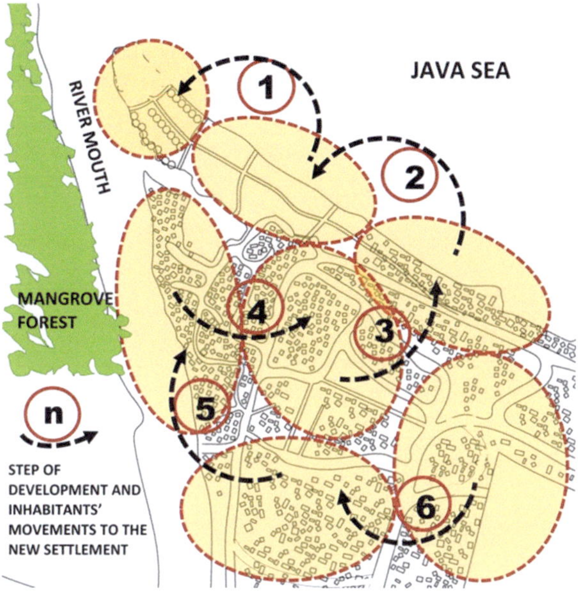

Fig. 10.10 Step by step development of Muara Angke new settlement

Fig. 10.11 Completed development of the new fishing settlement

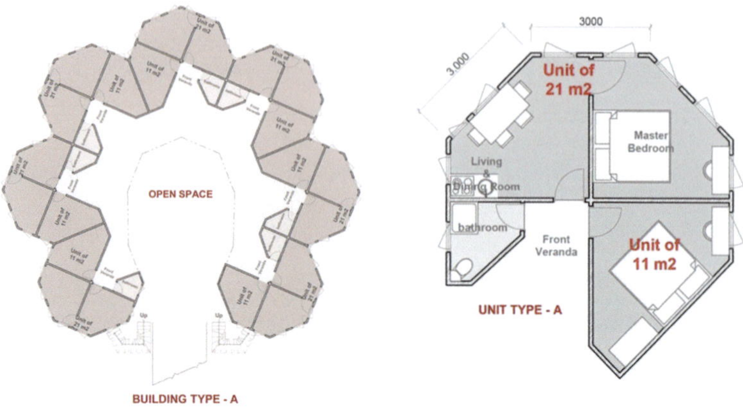

(a) (b)

Fig. 10.12 Floor plan design of building Type-A, consisting of 7 units Type-A (**a**). Each unit consists of two sub-units with floor areas of 21 and 11 m^2 (**b**)

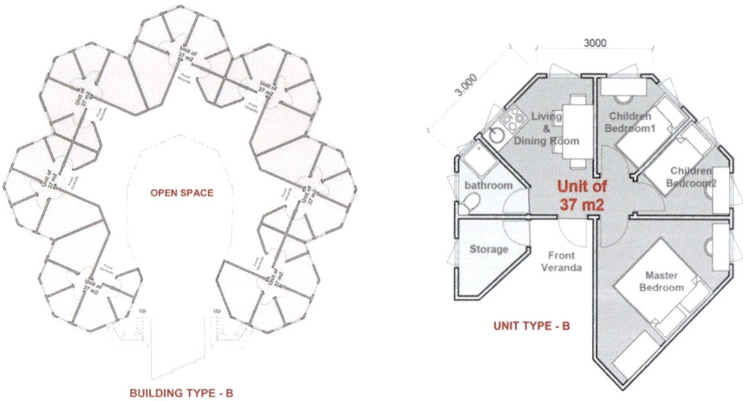

(a) (b)

Fig. 10.13 Floor plan design of building Type-B, consisting of 7 units Type-B (**a**) each with floor area of 37 m^2 (**b**)

5.3 Sustainable Concepts: Sanitation, Energy Generation and Food Gardens

The shape of the dwelling unit resembles a shell, while after being attached to form a building, the form of the building is likely to resemble a mushroom. It is expected that the shape of the building and its arrangement to form the new settlement would be natural and organic. Furthermore, the arrangement of the dwelling units is expected to support the existing close relationships amongst the inhabitants.

The new settlement is proposed to be sustainable. The first sustainability concept applied in this design is that it deals with a way to reduce the damaging effects of floods, both from the sea water and the overflow ofthe riverwater. Secondly, it provides better and proper sanitation for the whole collection of dwelling units. Clean water would be supplied both from the rainwater harvesting system and the Jakarta water company and distributed to all of the houses and also to the public toilets in the public area. The waste water would be treated safely before being discharged to the river nearby. Solid waste from the households will be collected and separated between inorganic and organic waste. The organic waste combined with the sewage waste is used to produce biogas (Moment 2011). Although, it may not be sufficient, all the gas produced by the generator is to be used for domestic cooking.

To generate its own electricity, solar photovoltaic panels (PV) will be installed in some parts of the roof. It has been calculated that 2–3 m^2 of solar panels mounted on the roof of each dwelling unit would be enough to supply the lighting demands for every single family. Two wind turbines are proposed to be constructed at the seaside of the north part of this settlement (Fig. 10.11), in order to supply the electricity demands for public purposes, such as outdoor lighting, refrigerators and some public areas and public buildings. In the early phase, there would be an electricity supply available from the state electricity grid.

There will be some gardens provided in this settlement to grow fruits and vegetables to meet some of the domestic demands.

References

Alikodra HS (1996) The impacts of land reclamation of Jakarta bay on Mangrove ecosystem. Media Informasi J 5(1):31–34, April 1996, Indonesia

Fong WYK (2014) Living and dying in Tai O: sustaining the heritage of stilt houses in the fishing village of Tai O, dissertation of Master of Science in Conservation, University of Hong Kong (pdf). Available at http://hub.hku.hk/bitstream/10722/208077/1/FullText.pdf?accept=1. Accessed 13 July 2016

Garuda (2016) The magazine of Garuda Indonesia. Available at http://v2.garudamagazine.com/features.php?id=125. Accessed 18 July 2016

Girardet H, Mendonca M (2009) A world renewable energy, ecology, equality a report for the world future council. Gren Books Ltd, Totnes

Housing-estate.com (2016) Rusun Muara Angke dan Cengkareng Barat Batal Dibangun (Flats of Muara Angke and West Cengkareng are canceled to be built). Available at http://www.housing-estate.com/read/2016/07/13/rusun-muara-angke-dan-cengkareng-barat-batal-dibangun/. Accessed 19 July 2016

Jakarta Climate Office (2016) Weather and tidal forecast at Tanjung Priok Harbor and Its Surroundings (pdf), available at http://kkp.go.id/assets/uploads/2016/01/12_Prak_pelabuhan_jar.pdf. Accessed 28 June 2016

Jakarta Raya (2016) Pemprov DKI Tetap Sediakan Rusun Untuk Warga Muara Angke (The Jakarta authority is still preparing to build multi-storey flats for Muara Angke inhabitants, E-Paper Bisnis Indonesia. Available at http://jakarta.bisnis.com/read/20160609/77/556193/pemprov-dki-tetap-sediakan-rusun-untuk-warga-muara-angke. Accessed 19 July 2016

Karyono TH (2004) Hunian Tanpa gusuran: Masalah Hunian Masyarakat Bawah di Kota (Housing without eviction: problems of low-income people live in the city), paper. In: Proceedings of National Conference on Dwellings for All People, Duta Wacana Christian University, Yogyakarta, 9 October 2004

Moment, Modern Water Management, newsletter 6 (2011) Available at http://www.rfkl.se/Documents/MOMENT/Newsletter/Newsletter_December.pdf. Accessed 20 July 2016

Patandianan MV, Toban Z (2011) Identifikasi Pengembangan Permukiman Nelayan oleh (Identification of fisihing settlement development by) Neighborhood Upgrading and Shelter Sector Project (NUSSP) (pdf), paper in Research proceeding, Faculty of Engineering, Hasanudin University, Makassar, Indonesia. Available at http://download.portalgaruda.org/article.php?article=143708&val=2170. Accessed 13 July 2016

Stocker TF et al (eds) IPCC (2013) Summary for policymakers. In: Climate change 2013: the physical science basis. Contribution of working group I to the Fifth assessment report of the intergovernmental panel on climate change. Cambridge University Press, Cambridge, MA/New York

Wikipedia, Fishing Village (2016) Available at https://en.wikipedia.org/wiki/Fishing_village. Accessed 13 July 2016

Chapter 11
Proposed Design of Bicycle Lanes Around Jakarta's East Flood Canal

Diah Anggraini, Sutrisnowati Odang, and Iman Mustadjab

Abstract Jakarta's transportation is dominated by motorized vehicles powered by fossil fuel. The city has a poor provision of sidewalks, discouraging people from walking even to reach a nearby destination. No bicycle lanes are provided in the city, minimizing the number of people riding bicycles in the city. This situation has been creating heavy air pollution in the capital city. In order to reduce the city's air pollution and to improve its environment, a number of Car Free Day (CFD) programmes have been launched by the city authority since a few years ago. This programme is reported to have helped in reducing the air pollution somewhat. Along with the development of Jakarta's 23.5 km long East Flood Canal, known as 'Kanal Banjir Timur' (KBT) in East Jakarta, the canal banks have also been developed to provide some public facilities, including bicycle lanes. However, the limited connection of the cycle lanes to the other parts of the city has discouraged people from using them. This study aims to establish an urban planning guideline to manage the development of a bicycle lane system in the city. The case study is taken in a selected area around the KBT in East Jakarta, In order to encourage people to take bicycles to work, this study tries to find out possible connection lanes between the KBT bicycle lanes and some of the public facilities in this area. Along with the improvement of the public transport system, adding some bicycle lanes and pedestrian short cuts between public facilities is likely to reduce the number of motorised vehicles in the city, thus improving the air quality of the city.

Keywords Air quality • Air pollution • Bicycle lane • Linkage • Motorized vehicle • Urban area

---

D. Anggraini (✉) • S. Odang • I. Mustadjab
School of Architecture, Tarumanagara University, Jakarta, Indonesia
e-mail: diah_ismono@yahoo.com

T.H. Karyono et al. (eds.), *Sustainable Building and Built Environments to Mitigate Climate Change in the Tropics*, DOI 10.1007/978-3-319-49601-6_11

1 Introduction

The increased number of motorized vehicles has deteriorated the air quality of a city like Jakarta. To respond to this problem, the Jakarta authority has promoted a Car Free Day (CFD) program in 2002. Research conducted by the Jakarta Regional Agency for Environmental Management (Badan Pengelolaan Lingkungan Hidup Daerah, BPLHD) showed that the reduction of CO, NO_x and dust pollution was noticeable as a consequence of the CFD program (Suryani 2015). In some developed-industrial countries, motorized vehicles system has been switched to some extent to non fossil fuel based transportation. People are encouraged to use non-motorized modes, such as bicycle and foot. Cycling is preferable in a number of European towns and cities as it is cheaper, healthier and eco-friendly. Unfortunately, most Indonesian cities are not prepared with their masterplan design to incorporate cycling as part of the city's transportation system. Bicycle lanes are not provided along the roads and it appears impossible to add them onto the existing roads.

Along with the KBT plans, the Jakarta authority has planned the canal side area as part of the city's green and public spaces, including bicycle lanes. Green spaces are built along the canal, while other spaces are used for other purposes, such as an inspection road for cleaning the canal, jogging tracks and the 6.5 km bicycle lanes (Adhi 2010). The KBT public spaces tend to reach their peak function during the weekends and holidays only, while on the weekdays they are relatively unused. During the weekdays, the bicycle lanes tend to be occupied mostly by motorized vehicles, not by bicycles. The provision of this lane to encourage people to use bicycles seemed unsuccessful. A possible reason for this is that the existing bicycle lanes are not connected directly to the residential areas nor integrated to the other parts of the city. Moreover, the bicycle lanes are discontinued when crossing the other roads (Anggraini 2012).

The provision of bicycle lanes needs to be synchronized with the city infrastructure and to its system for increasing common interest for utilization of non fossil-fuel based transportation. One of the aims of this study is to give a recommendation for the Jakarta authority to improve the city's guidelines regarding bicycle lane plans.

The proposed area to be developed is a part of the bicycle lane, stretching from Cipinang Besar Selatan to Pondok Kopi, however, a field visit to observe the actual condition was made only in a particular area around Duren Sawit to Dharma Persada University campus at Pondok Kelapa stretching about 2.3 km long (Fig. 11.1). Observation has made on the north side of the bicycle lane since the south side was likely to be less crowded. Observations have been made on the following aspects:

- Physical conditions of the bicycle lane and its facilities, cyclists' number and profile.
- Movement pattern of people from and to the residential area around the KBT;
- Connections of roads between the residential areas and the KBT bicycle lanes

Fig. 11.1 Location of study

2 Motorised Vehicles and the Air Quality of the City

Due to its direct contribution to health and comfort, air quality of the city is a fundamental environmental element to be considered. According to Jakarta's BPLHD (2011) motorized vehicles contribute 27 million (62.2%) tonnes NO_x per year and 590 thousand (25,8%) tonnes CO per year; while industrial-related activities were responsible for the emission of other pollutants, such as SO_2. According to Jakarta Statistic Bureau (2015), in 2013, there were 16 million motorized vehicles, including motorcycle in Jakarta. Amongst these vehicles, 74% were motorcycles, 18% were private cars and the rest were public transport and cargo cars (Jakarta statistc Bureau, 2015). The total length of roads in Jakarta was 7.650 km, covering an area of 40.1 km^2 or about 6.2% of Jakarta's area. The number of motorized vehicles has grown significally much faster than the extension of the new roads, creating heavy traffic congestion anywhere in Jakarta. This circumstance has also contributed to higher emissions, deteriorating Jakarta's air quality. It was stated by the Jakarta BPLHD that Jakarta's air quality has been worsening. In 2005–2006, there were 50 days a year that the Jakarta's air quality was unacceptable and it has worsened every year since then (BPLHD 2011).

A research project conducted in Surabaya (Kanaf and Razif 2010) showed that CFD program has reduced the air pollution in the city. Jakarta's CFD program was initially held in 2008–2011 in five districts and has contributed a significant reduction in air pollution from car emissions (Suryani 2015). By considering the above facts, shifting the transport modes from motorized to non motorized vehicles would reduce the city's air pollution. Cities around the world have been redeveloped to provide more spaces for non-motorized vehicles and public transport. In 1988, Ashville, in North Carolina USA, reconfigured the city by making such comphrehensive changes in its master plan. Taking into account the existing physical city conditions, a bicycle traffic plan has been developed to reduce traffic congestion in the city of Ashville and to improve its environment (DRAFT 2005).

3 Bicycle Lane as a Linkage

The atmosphere of the city is dependent on the interaction between activities and functions. Pedestrian paths and public transport connect spaces with different functions, creating some activities in between and mixing people with various social and cultural backgrounds (Wiryomartono 2002). Linkage is the connectivity between places in the city. Linkage holds the unity of the city; it helps people to understand the overall plan of the city. Zahnd (1999) categorizes linkage into three types: visual, structural and collective. Linkage connects nodes to serve movements from one point to other points to establish a series of activities, such as commercial, educational, recreational and residential. Cycling as a non-motorized mode of transport enables people to carry out those activities.

Japan and Singapore can be good examples of cities which have successfully operated bicycle lanes. In Japan, 14% of journeys are done by bicycle. People use bicycles to get the 'park and ride' areas, changing and continuing the journey by public transport (Kompas 2012). Singapore has promoted a transportsystem, the Park Connection Networks (PCN) (Opini 2010), which includes a green corridor that connect gardens, squares and parks to the other parts of the city, including MRT stations, bus stops, markets and other commercial facilities.

4 Observation Around Jakarta's East Flood Canal (KBT)

The field observation carried out for this research covers several aspects, including: – physical conditions of the bicycle lanes and the cyclists' profile at KBT; Residential buildings' sizes and their structural forms around the KBT; residents daily activities, such as the times they go to offices, to schools, shopping, etc., since potential cyclists live in these areas (Anggraini 2012). Possible accessibilities of cyclists riding the streets around KBT to the KBT bicycle lanes. Figures 11.1, 11.2, 11.3, 11.4, 11.5 and 11.6 show the area of observations and things recorded during the observation.

Figures 11.7 and 11.8 show the possible traffic connections around the KBT and the KBT bicycle lanes. One of the main roads which is connected to the KBT area is Colonel Sugiono Street, located between Pahlawan Revolusi Street and Raden Inten Street. This street has eight spots for entry and exit to KBT, where the first and the eighth spots form the main access. These streets are considered as primary and secondary arteries with quite heavy traffic. No road separation and pedestrian

Fig. 11.2 Area of study

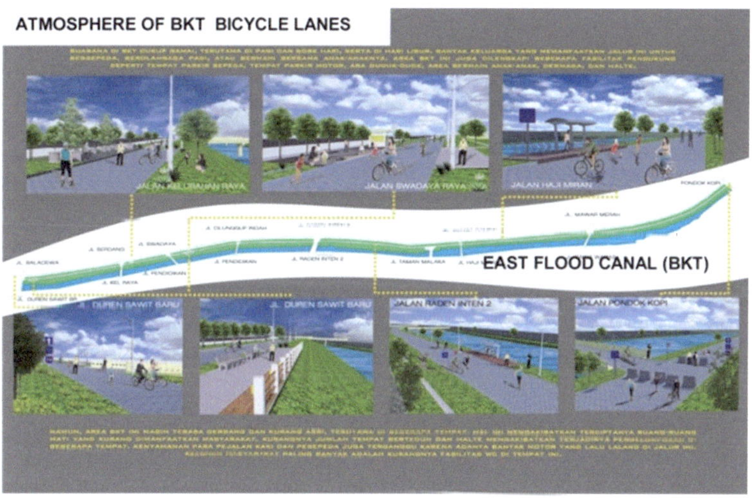

Fig. 11.3 Design images of KBT bicycle lane

crossing are provided on these streets, creating problems for pedestrians and cyclists as they are mixed up with motorised vehicles.

The traffic in these streets is dominated by private cars and public transport, with various speeds from slow to fast. Similar types of transport modes are found on the smaller roads in this area. On these roads pedestrians are mixed up with cyclists, street vendors and other vehicles.

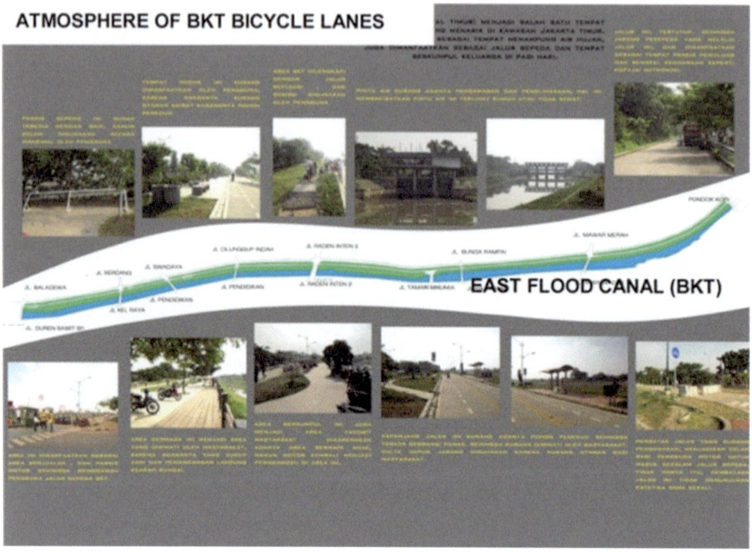

Fig. 11.4 View of completed KBT bicycle lane area

Fig. 11.5 Activities around the KBT bicycle lanes

5 Observations Regarding Public and Social Facilities

Figures 11.9, 11.10, 11.11 and 11.12 show the various types of facilities in this area, schools, shops, offices, health care centres, churches and mosques. These facilities are attracting people to come and go, generating traffic around this area.

In terms of educational facilities, primary and secondary schools are located less than 1 km from the KBT, while the high school is around 3 km away from KBT. Those schools are within the walking distance from KBT. Figure 11.9 shows that schools are grouped within the yellow areas. Other than taking a walk, students would be encouraged to use bicycles if these schools were connected with bicycle lanes.

In terms of health care facilities, Fig. 11.10 shows the location of hospitals, health care centres and chemists. Ther are a general hospital (Duren Sawit Hospital) and two maternity hospitals around the KBT. Some additional health care centres

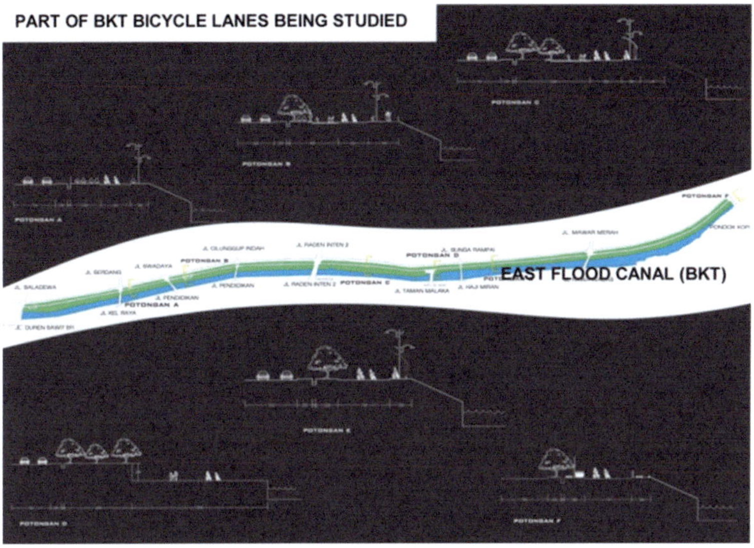

Fig. 11.6 Sections of KBT bicycle lane

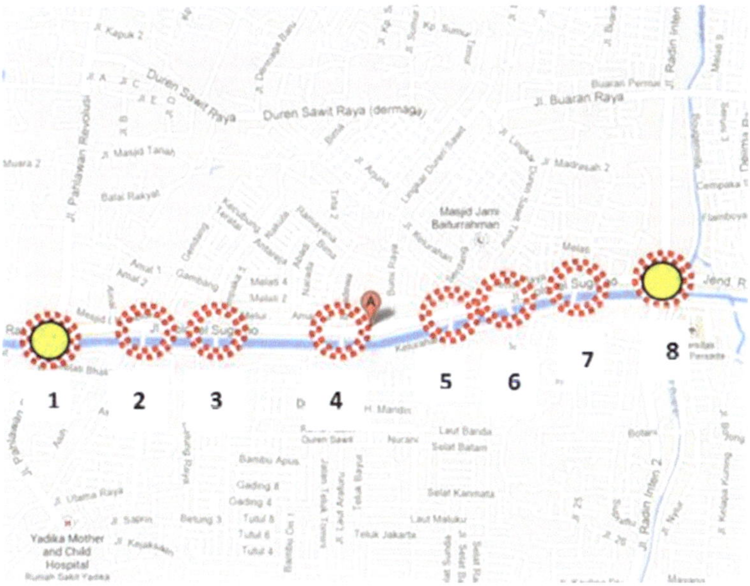

Fig. 11.7 Bicycle lanes: Entry and exit points at Colonel Sugiono street

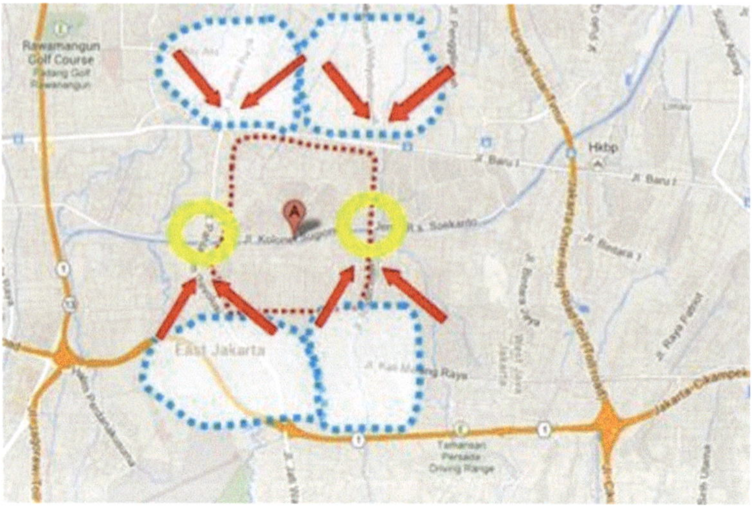

Fig. 11.8 Bicycle lanes direction from North and South via point 1 and 8

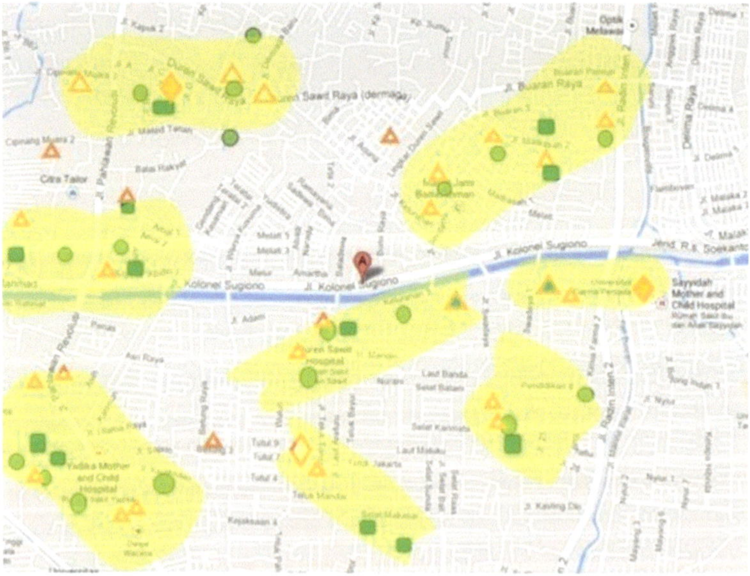

Fig. 11.9 Educational facilities

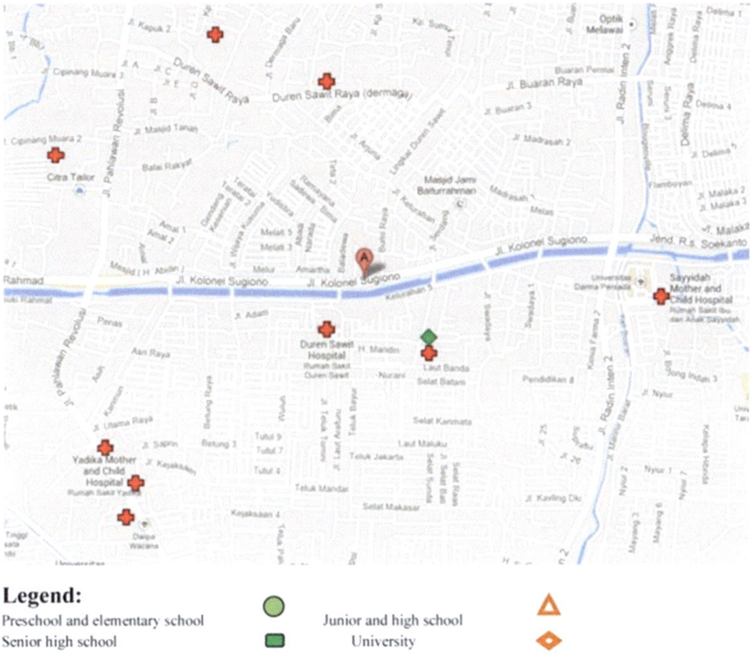

Fig. 11.10 Location of hospitals, health care centres and chemists

and chemists are located around the residential areas, less than 3 km away from the KBT area.

Figure 11.11 shows the location of offices, shops, shophouses, restaurants and cafés, which are around the main roads, while markets are situated at the secondary roads. There are residential markets located around the residential areas. Most of the markets are located a distance of no more than 2 km away from the housing complexes, which are within the walking and cyling distances.

Figure 11.12 shows the location of religious facilities, such as Mosques and Churches. The locations of these buildings are various. A number of Mosques are located at the main roads, however, most of them are found in the residential areas. The Churches, on the other hand, tend to be located on the main or secondary roads, only few are in the residential areas.

The provision of bicycle lanes is likely to stimulate people to use bicycles than motorcycles, as they are light and enjoyable as a means of transport. Curently the provision of KBT bicycle lanes is still limited to bike-fun activity only, particularly during the week end and holiday as the lanes are limited in their connections. Impoving the bicycle lanes' conections to other urban facilities and residential areas as proposed in this study, bicycle lanes would encourage people to use more bicycles than motorized vehicles, particularly to travel short to medium distances. The

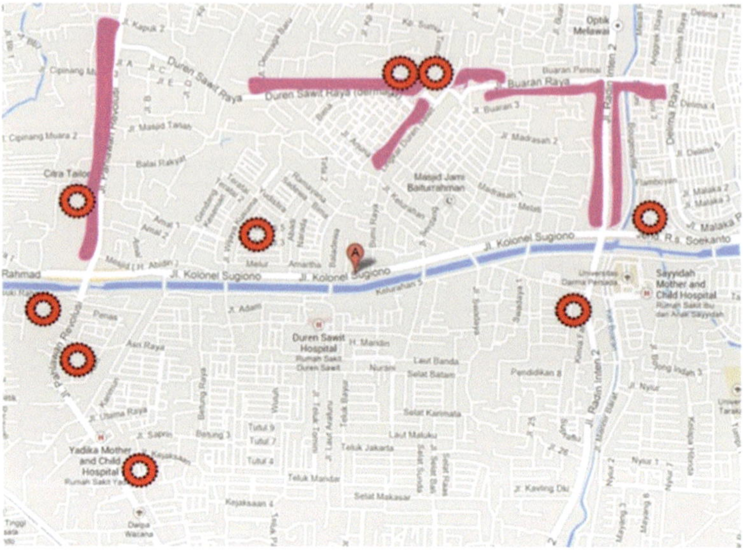

Fig. 11.11 Location of offices, commercial, trading facilities and markets

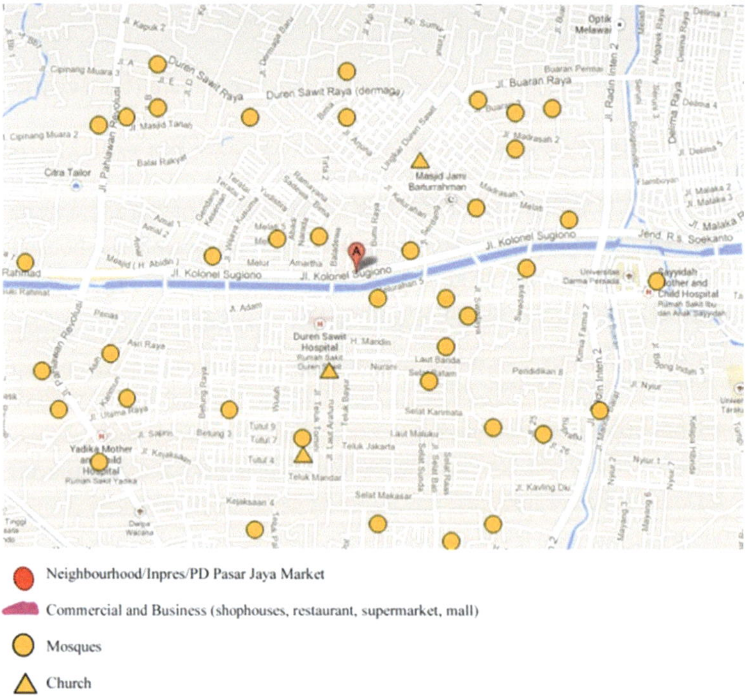

Fig. 11.12 Religious facilities

less motorized vehicles are used in the city, the less carbon dioxide is emitted, reducing city air pollution and greenhouse gases.

6 The Proposed Bicyle Lanes Design as an Urban Linkage

The planning for improved linkage of the KBT bicycle lanes focuses on connecting meeting points and magnets. The connectors (linkages) are planned as connections on the basis of the daily activities, going through facilities which are located in different groups such as school, market, hospital, health care centre and religious facilities. There are three types of connectors described below. These connectors provide access to the surrounding area, to the secondary roads, where all the facilities are located. Cyclists are encouraged to utilize the tracks toward intended facilities, where every track is designed according to the existing characteristics.

Figure 11.13 shows the whole proposed system of bicycle lanes around the KBT and the way they are connected to each other. There are three type of connectors (linkages) each provided with a slightly different design and function, as follows:

1. Main connectors (red colour) are provided on the road with signed bicycle lanes (SNI 2002). The tracks are marked on the road with a green colour, called green

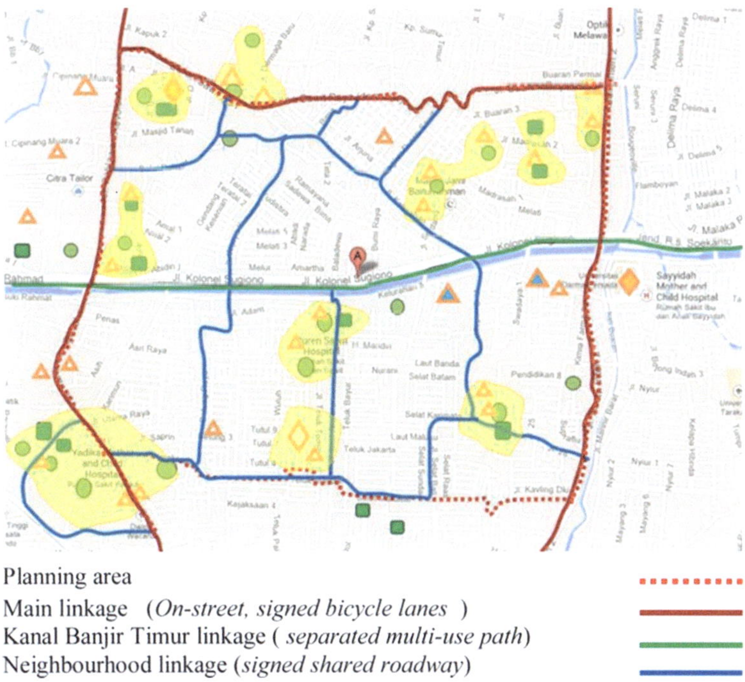

Fig. 11.13 Design proposal for bicyle routes around KBT

carpet. This track is a mixed-use, which accommodates all types of transport modes, including motorcycles.

2. KBT connector (green colour) is the existing KBT bicycle lane. It has its own track, separated from the road. This link is intended for use only by non–motorized vehicles, particularly for bicycles. Along with this track, a pedestrian path is provided at the side with a permanent separator.
3. Neighbourhood connectors (blue colour) are located around the residential areas. These connectors connect cyclists to various facilities around this area and to the KBT bicycle lanes. This linkage uses a signed shared roadway system, where roads are marked with bicycle signage for cyclists' safety purposes.

Figures 11.14, 11.15, 11.16, 11.17, 11.18 and 11.19 show details of the proposed bicycle lanes design at the intersections with existing roads. The green colour indicates an area used for mixed vehicles. The yellow colour indicates areas of bicycle lanes on the main road side, with no separator, while the purple colour indicates areas of bicycle lanes with a permanent separator.

The provision of appropriate bicycle lanes in urban and residential areas, such as shown here in Jakarta, would encourage people to change their habits to use more non-motorized than motorized vehicles. The more people are using bicycles rather than cars and motorcycles would help to reduce carbon emissions of the city, thus, helping the air of the city to be cleaner and at the same time reducing global warming and climate change.

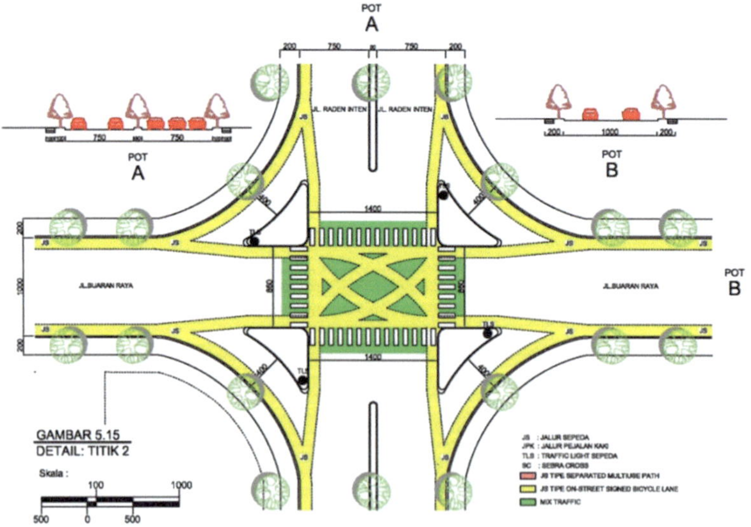

Fig. 11.14 Detail-1: Circulation design of bicycle lanes at the crossroads

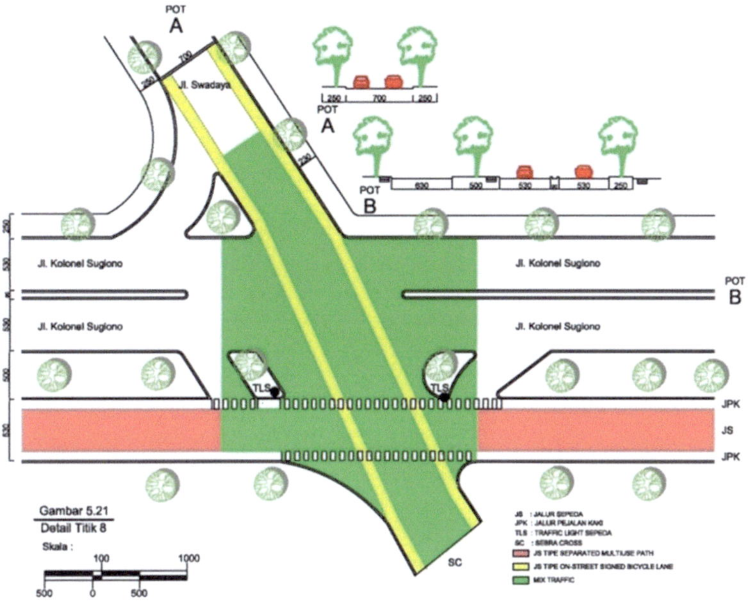

Fig. 11.15 Detail-2: Circulation design of bicycle lanes when crossing the roads

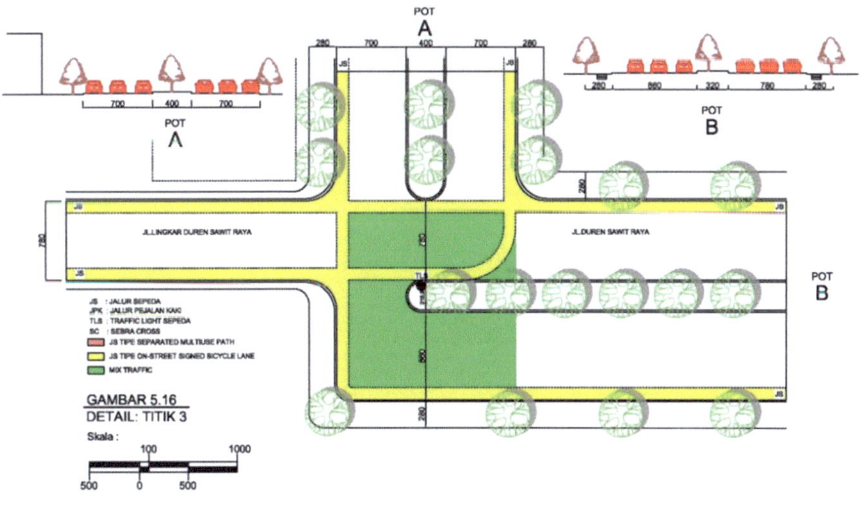

Fig. 11.16 Detail-3: Circulation design of bicycle lanes at a particular junction (a)

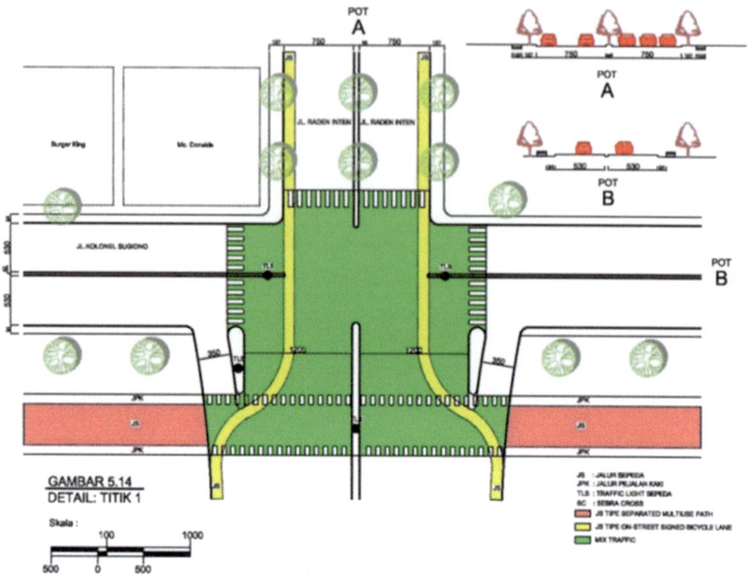

Fig. 11.17 Detail-4: Circulation design of bicycle lanes at a particular junction (b)

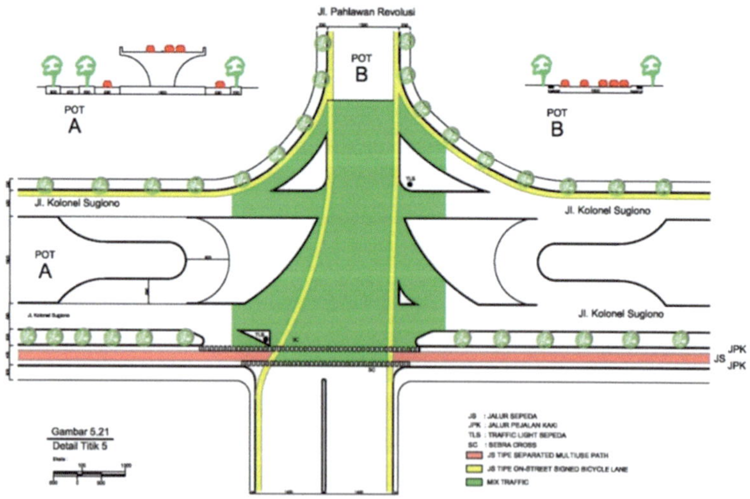

Fig. 11.18 Detail-5: Circulation design of bicycle lanes at a particular junction (c)

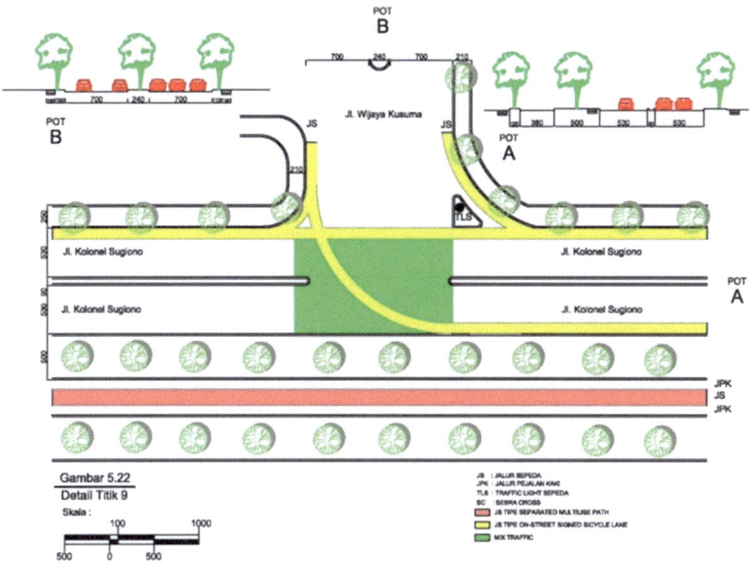

Fig. 11.19 Detail-6: Circulation design of bicycle lanes at a particular junction (d)

References

Adhi RK (2010) Banjir kanal Timur, Karya anak bangsa. Grasindo-Penerbit PT Gramedia Indonesia, Jakarta, pp 217–221

Anggraini D (2012) Jalur Sepeda Dalam Konteks Tautan Antar Kegiatan di Perkotaan. In: Proceedings of International Conference: *sustainable built enivironment in the tropics: new technology, new behavior?* School of Architecture, Tarumanagara University, Jakarta, Indonesia. 12–13 November 2012

Badan Pusat Statistik Propinsi DKI Jakarta (2015) Transportation statistic of Jakarta 2015. http://jakarta.bps.go.id/backend/pdf_publikasi/Statistik-Transportasi-DKI-Jakarta-2015.pdf. Accessed 22 Aug 2016

BPLHD (2011) Laporan status Lingkungan Badan Pengelolaan Lingkungan Hidup Daerah

DRAFT (2005) City of Ashville Pedestrian Plan Update. http:www.ashvillenc.gov/Departments/Transportation/TrafficEngine

Kanaf and Razif (2010) The efficiency of car free day program to the reduction of carbo emissions. ITS Surabaya

Kompas (2012) Sepeda Pilihan Sehat Warga Tokyo (Bicycle, The Healthy Choice for Tokyo Citizens). 4 October 2012

Opini (2010) Park connector network: Konsep Jalur Transportasi non-motorized dan Penghijauan Kota. 29 April 2010

SNI (2002) Standar Nasional Indonesia-SNI-1733-2002

Suryani R (2015) Five Years Car Free Day in Jakarta. Better Air Quality Conference 2012, Hongkong, 5–7 December 2012

Wiryomartono BP (2002) Urbanitas dan Seni Bina Perkotaan. hal 163. Balai Pustaka. P, Jakarta

Zahnd M (1999) Perancangan Kota Secara Terpadu. Teori Perancangan Kota dan Penerapannya:(hal 107, 108). Penerbit Kanisius dan Soegijapranata University Press, Yogyakarta

Chapter 12
Flood Responsive Design of the Low-Income Settlements in Kampung Melayu, Jakarta, Indonesia

Tri Harso Karyono, Nanda Hidayah Melyan, Siti Yaumilia Salsa, and Elinda Fariz

Abstract The coastal city of Jakarta experiences frequent floods, particularly during the rainy season. The main areas which suffer from the floods are those on the riverbanks. The Ciliwung river, flowing in the middle of this capital city, is likely to be the main cause of flooding when its upstream water flows through the capital during the rainy season. Deforestation in the headwater regions due to the development of farms, plantations and housing has reduced the absorption of the rainwater, sparking massive water flows to the lower areas like Jakarta, creating floods in this city. One of the areas which has the most severe problems due to the annual floods is Kampung Melayu. This area is mainly occupied by low income settlements, in which a proportion of the inhabitants are people coming from areas outside Jakarta, such as West Java, Central Java and the other parts of Indonesia. Due to the frequent floods, some of the inhabitants have tried to reduce the impact of flooding by raising the floors of their houses. However, this has not been enough as the river water levels tend to rise. This study tries to compile data of the inhabitants in this settlement, looks at the efforts the people have made to reduce the impact of the floods and finds appropriate design solutions to overcome the problem.

Keywords Design solution • Floods • Squatters • Stilt houses • Sustainable settlement

1 Introduction

In developing countries such as Indonesia, there are a proportion of people living as squatters, inhabiting inappropriate houses in unsuitable locations, such as at the sides of railway tracks and also along riverbanks. Some of these squatters face the

T.H. Karyono (✉) • N.H. Melyan • S.Y. Salsa • E. Fariz
School of Architecture, Tanri Abeng University, Jakarta, Indonesia
e-mail: t_karyono@yahoo.com

T.H. Karyono et al. (eds.), *Sustainable Building and Built Environments to Mitigate Climate Change in the Tropics*, DOI 10.1007/978-3-319-49601-6_12

problem of annual floods, occurring during the rainy season. This kind of settlement is vulnerable to being evicted as it is considered to be illegal as it occupies land owned by the government (Karyono 2004). The areas used for this kind of settlement are supposed to be for green areas and public facilities.

People who live in these areas mostly have low incomes. They cannot afford to live in a more appropriate place, which is considered to be legal. The settlement tends to be in the form of shanty houses, located close to the railway tracks or riverbanks. Most of the people living in this type of settlement are migrants, coming from small towns or villages outside Jakarta. The migration which is more popular as urbanization increases is normally caused by the economic pressure in migrants' hometowns. They cannot get the expected jobs in their hometowns and villages, while in the big cities, such as Jakarta, there are more opportunities to get jobs to improve their lives. The cause of urbanization in Indonesia tends to be different compared to the other developing countries. In Indonesia, movement from villages to big cities is not directly related to industrial development as in most of the developing countries. A survey in 1995 showed that 59.6 % of migrants from rural areas had moved to the capital to seek employment (McCarthy 2003). However, based on regular discussions with a number of migrants regarding the reasons for their moving to the capital city, the reason was that a lot of things could be done in the capital to earn money. A high expectation to get a better life is likely to be the main reason why the migrants are moving to the big cities such as Jakarta. Their choices are an indirect response to industrial development. They are not expecting to work in the factories, but rather to work in the informal sector, such as house helpers, street vendors, drivers and construction workers.

This study aims to provide a way to solve the current problems of squatters occupying the Jakarta riverbanks which are vulnerable to the floods. In most of the big cities, the physical conditions of the squatters on the riverbanks tend to be similar, mostly they are located in the area between rivers and roads.

There are a number of squatter settlements in Jakarta, however, some of the most vulnerable settlements are those located in the Kampung Melayu subdistrict in the central area of Jakarta. One of the settlements in this area is taken as a case study to see the actual problems and find out the solution. Although this case study is taken at a Kampung Melayu settlement, the proposed design solution for this settlement is likely applicable for others which have a similar condition.

2 History of the Riverbank Settlements in Jakarta and the Problem

Most of all the settlements built on the riverbanks tend to be illegal. The riverbanks are areas belonging to the government and are expected to be used for green areas and public facilities, including service roads to clean the rivers. Figure 12.1 shows the empty land along the Ciliwung river.

Fig. 12.1 An area of a riverbank between a river and a road was empty before the urbanization process, as in Kampung Melayu

Due to the urbanization process, the migrants, dominated by low income people from outside Jakarta, moved to the capital and started to live on some empty land areas, including riverbanks in the central area of the city. A decade ago, about 20–25 % of Jakarta residents lived in kampungs (urban villages), with a further 4–5 % who resided illegally on riverbanks and other empty lots (McCarthy 2003).

Some of the migrants were jobless when they arrived in Jakarta. Since they need a place to live and to earn money, they started to build temporary shelters, using waste materials like cartons, plastic and wooden boards. They use the shelters not only as dwellings, but also as workplaces and places from where they can sell things to earn money. Figure 12.2 shows how a number of migrants started to build shelters on the riverbank.

Along with the improvement of their daily income from the informal business, step by step, the shelters have been improved to be more permanent, using bricks, wood and bamboos. In the past, the local government was likely to ignore the development of these squatter settlements, which, in fact, occupied the government land.

Since there had been no action taken by the local authority to stop them, more people started to come and occupied more spaces in this area. They built more shelters to accommodate their families and relatives (Fig. 12.3). The empty land of riverbanks started to become a squatter settlement or uncontrolled settlement. The settlements have grown massively in many places in the central area of Jakarta (Fig. 12.4). Many of the shelters in these kinds of settlements have been improved and built permanently using modern building materials such as bricks, corrugated aluminium, steel sheets and timber.

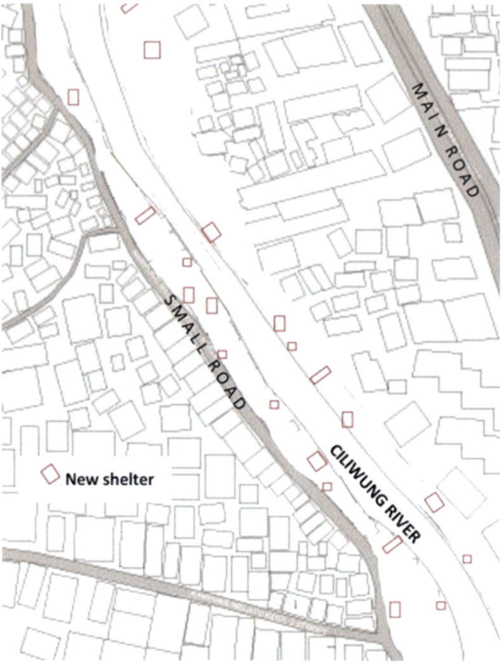

Fig. 12.2 In the beginning of the urbanization process, migrants started to build shelters on the riverbank as in Kampung Melayu

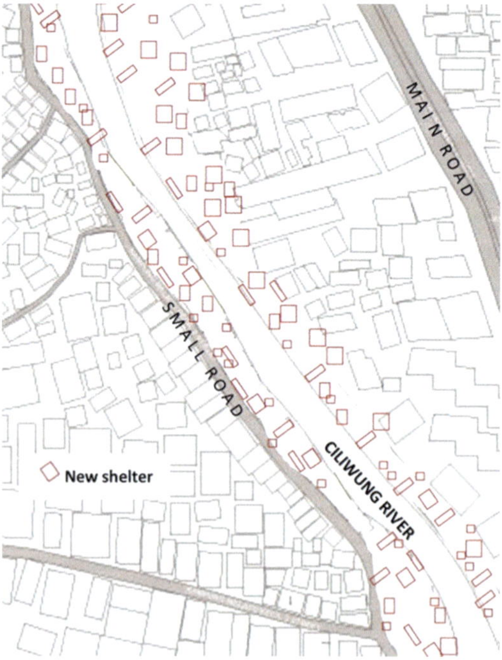

Fig. 12.3 More shelters are built on the empty spaces of the riverbank as in Kampung Melayu

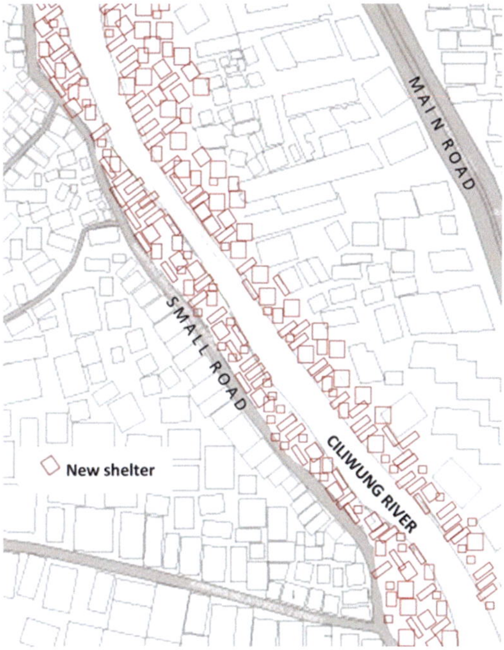

Fig. 12.4 Every available space on the riverbank has been occupied, creating a dense squatter settlement as in Kampung Melayu

These kinds of settlements have survived for decades for different generations. The inhabitants are likely to be unaware that they live in shelters which are considered to be illegal. They always think that their shelters are legal as they can get access to electricity from a state electricity company. Moreover, some of them have installed landline telephones provided by the state telephone company (Karyono 2003). These can be seen as proof of their legal existence in this area. Some of them even bought and sold their shelters. A number of low income people are likely to be attracted to live in this kind of settlement due to its strategic location in the central area of the city.

With the problems of floods occurring annually in these kinds of settlements, the local authority started to take some actions to overcome this problem. The authority believed that the shelters covering the surface area of the riverbanks is one of the factors that is worsening the floods. They determined that the shelters should be removed and replaced with green areas to absorb the overflowing river water during the rainy season. By moving out people from these areas and clearing away all the buildings, the authority thinks that they would have triple benefits. Firstly, this action would help the inhabitants to be rescued from the annual floods; secondly, by removing all the buildings on the riverbanks, more land surface would be available to absorb more water during the times of excess flow, thus, it would lower the water levels during the floods; and thirdly, it would improve the scenery of the town landscape as the riverbanks would be seen as greener with trees covering this area.

Based on these reasons, the local authority has decided to evict all the squatter settlements in Jakarta. Step by step, they started to warn the inhabitants to leave the settlement voluntarily. The authority has promised that all the evicted families and people would be relocated in low cost housing flats already built by the authority.

However, instead of accepting the authorities' initiatives, the inhabitants tend to reject the plan and resist moving. The most common reason that most of the inhabitants reject this plan is that the new flats are usually located too far from their current settlements. Should they accept the government plan, they will be facing more problems. Most of the inhabitants work close to their settlements, some are street vendors, selling foods and other stuffs nearby to their current shelters. Moving away from the current settlements would lose their jobs.

Furthermore, they will also have problems with their children. Most of the kids have enrolled in schools nearby their current settlements. Moving away from the current settlements would cause the kids to have to quit their current schools and enroll in the other schools close to the new flats. Rescuing the inhabitants from the floods would be offset by some new problems, such as losing jobs and having to move the children to new schools. The other important reason for the rejection is that the inhabitants have to pay the rent for the flats, while they currently live rent-free in their current houses. The monthly expense for the flat could reach one third of their monthly incomes, including the electricity and water bills, while staying in the current houses, they only spend about one fifteenth of their monthly income for the electricity and water. Moving will make them significantly poorer.

3 Kampung Melayu Squatter Settlements

One of the riverbank settlements which is considered to be illegal and having frequent floods is the group of squatters in Kelurahan (subdistrict) Kampung Melayu, in East Jakarta. Literally, Kampung Melayu means Malay kampong or Malay town-village. This area had been inhabited by Malay descendants from Southern Thailand in the seventeenth century during the Dutch occupation period. However, today, the literal meaning of Kampung Melayu does not reflect the actual ethnic make-up of people living in this settlement. This area has been inhabited by multi-ethnic Indonesian people since the country declared its independence from the Dutch in 1945. The size of the Kampung Melayu subdistrict is 47.83 Ha, with a population of about 30,000 in 2012 (Widianto 2012). This subdistrict has eight RW (rukun warga) or Community Units, divided into 112 RT (rukun tetangga) or Neighbourhood Units (Widianto 2012). Figure 12.5 shows a part of Kampung Melayu squatter which has been evicted and has been built a new road.

The Kampung Melayu squatter settlement is situated on the Ciliwung riverbank right in the central area of the capital city of Jakarta. This settlement has suffered from annual floods, resulting from large amounts of water from upstream rivers in West Java that flow through this area during the rainy season. The Ciliwung river is one of the city's waterways most affected by floods. As a part of the Jakarta authority programme some of residents of this settlement have been evicted recently.

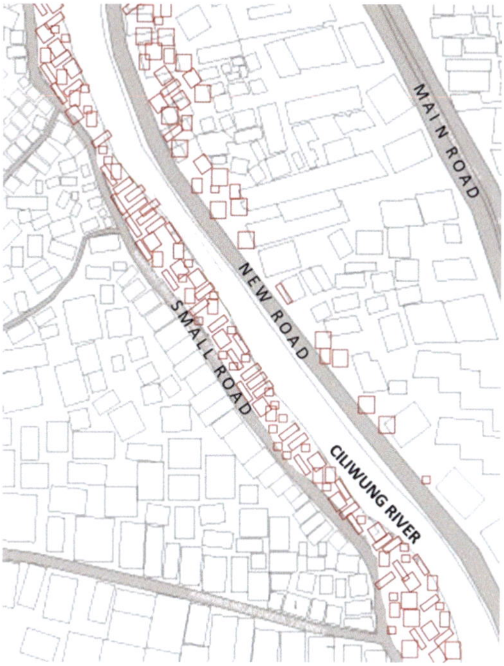

Fig. 12.5 A part of Kampung Melayu squatter at the east side of Ciliwung riverbank which has been evicted and has been built a new road

The area of Kampung Melayu is only about 2–3 m above sea level; with its climate similar to other locations in Jakarta. The Jakarta average temperature is around 28.5 °C, with a range of between 24 and 33 °C. The relative humidity is between 50 and 92 %, with an average of 78 % (BPS-Statistics DKI Jakarta 2006), while the precipitation is between 29.0 mm/month in the dry season (May to October) and 385 mm/month in the rainy season (November–April) (World Meteorological Organization 2016).

On 20th August 2015, the Jakarta authority took action to evict a major part of the Kampung Melayu squatter settlements. Even though there was resistance from the inhabitants, the authority has successfully forced them out of this area and moved them to the nearby flats belonging to the Jakarta authority. There could be a plan from the authority to move further to evict more inhabitants from this riverbank (Figs. 12.6, 12.7, and 12.8).

Fig. 12.6 A spot of the remaining Kampung Melayu squatter on Ciliwung riverbank in a normal situation with no flood

Fig. 12.7 Kampung Melayu squatter settlements are vulnerable during the floods

Fig. 12.8 Kampung Melayu shelters built with light materials

4 Profile of the Inhabitants and Problems Caused by Frequent Floods

Frequent floods take place in a number of areas in the capital, mostly every year. The causes of the floods are various (Karyono 2012), although the most severe flooding is likely caused by overflowing river water as occurs annually in Kampung Melayu.

In order to see the backgrounds and the problems of the inhabitants who occupied the Kampung Melayu squatter settlement, we conducted a field survey by visiting one Neighbourhood Unit (RT). There are a number of RTs in this settlement in which the profile of the inhabitants in the whole settlement tends to be similar. Therefore, the data collected in this survey tend to reflect the whole population of this settlement. The survey had been done a few months before the eviction took place on 20 August 2015. Using face to face interviews and helped by questionnaire checklists, data related to the inhabitants' backgrounds have been collected, including data on house sizes and the efforts of the inhabitants to reduce the impact of the floods.

Sixty four households, consisting of 29 males and 35 females, have been visited and interviewed during the survey. Two of the respondents were 20 years old, 10 respondents were between 21 and 30, 26 respondents were between 31 and 40 years old, while the remaining 26 respondents were over 40 years old.

In terms of their ethnic backgrounds, two respondents came from Banten, 33 respondents originated from Jakarta, 19 respondents from West Java, three respon-

Table 12.1 Respondents' occupations

Occupation	Number of respondents
Housewife	31
Self-employed	21
Motorcycle taxi driver (Ojek)	5
Daily wage labourers	4
Company worker	2
Public transport driver	1

Table 12.2 Respondents' family members

No of family members	Number of respondents
1–2	6
3–5	44
6–8	10
More than 9	4

dents from East Java, two respondents from Sumatera and one respondent from Kalimantan.

Table 12.1 shows the occupancies of the respondents collected in the survey. Since the surveys were conducted during the daytime hours, the category of housewife dominated the number of respondents. Out of 64 respondents, 31 of them were housewives. Self-employed was in second place, consisting of 21 respondents. The self-employed respondents were mostly scavengers, working as collectors of waste materials, such as plastic, glass and scrap materials from garbage dumps, also some of them run their own small shops in their houses. Five respondents were motorcycle taxi drivers, four respondents were daily wage labourers, two respondents were workers in small companies and one respondent was a public transport driver.

Table 12.2 shows the number of respondents' family members. They were six respondents with family members of two (husbands and wives) or a single person, 44 respondents with family members between three and five, ten respondents with family members between six and eight, while the remaining four respondents had more than eight members in the family, including their relatives.

In terms of the length of stay in their houses, four respondents had been staying less than 5 years, four respondents had been staying between 6 and 10 years, 55 respondents had been in their houses between 11 and 30 years and one respondent could not identify the length of the stay.

Table 12.3 shows the size of respondents' houses. There were 45 respondents who lived in houses of between 6 and 20 m^2, eleven respondents lived in houses of between 21 and 40 m^2, three respondents occupied houses of between 40 and 60 m^2, two respondents had houses of between 61 and 90 m^2, six respondents lived in houses of between 91 and 120 m^2, while the remaining two respondents occupied houses of more than 121 m^2. The house sizes ranged between 6 m^2, and 225 m^2, with

Table 12.3 Respondents' house sizes

Size of the house (m^2)	Number of respondents
6–20	45
21–40	11
40–60	3
61–90	2
91–120	6
More than 121	2

Table 12.4 Number of times respondents raised the floors throughout the time they lived in this settlement

No of times floors raised	Number of respondents
Never	44
1	12
2	5
More than 2 times	3

Table 12.5 Average number of floods experienced by the respondents annually

Number of floods	Number of respondents
1	9
2–5	43
Mostly every heavy rain	12

an average of 31.25 m^2. Those with a size of more than 60 m^2 tended to be occupied by more than one family.

Table 12.4 shows the number of respondents who had been raising the floors of their houses throughout the time they lived in this settlement, in order to try to prevent them from being flooded. There were 44 respondents that had never raised the floors, 12 respondents had raised once, five respondents had done it twice and three respondents had raised the floor more than two times.

Table 12.5 shows the average number of floodings of their houses that the respondents had been experiencing annually. There were nine respondents who had experienced a flood once in a year, 43 respondents had experienced two to five annual floodings and 12 respondents said they were likely to have floods whenever heavy rains took place.

Table 12.6 shows the high water levels which occurred during the floods that the respondents had experienced during their stays in this settlement. There were 57 respondents who had experienced flooding with water more than 2 m deep and only seven respondents who had less than 2 m during the floods. Overall, most of the respondents had experienced having to be evacuated during the floods. Those who had less than 2 m depth of water during the flood were situated on the upper sides

Table 12.6 The high water levels which occurred during the floods

The highest water depth (m)	Number of respondents
Less than 2	7
Higher than 2	57

of the riverbank just next to the road, so the small road can be seen to have less than 2 m depth during the floods.

Although there have been more floods occurring quite recently, the inhabitants are reluctant to leave their houses. They reject the Jakarta authority's plan to be reallocated to the new flats provided by the authority. The reasons for the rejection are, firstly, they believe that they live in a legal settlement as they have been living for a reasonably long time in this area and even subscribing to the state electricity grid. Secondly, they want to keep their current jobs and to keep their kids in their current schools.

5 Flood Responsive Design

Should the Jakarta authority take further action to evict more inhabitants in this area, there must be a careful consideration to be taken. Just moving people away from their current settlement is not likely to be the best solution to overcome the problems with floods. Leaving the current squatters and settling down in the new flats, away from their current shelters would create other problems for people, such as losing their current jobs, enrolling their children in new schools and spending more money for their living costs. The Jakarta authority has neglected the fact that the inhabitants are poor with limited income to rent flats. Furthermore, the inhabitants have established solid community ties between each other. Separating them from each other might not be an easy task.

5.1 Location and the Construction Process of the New Settlement

Based on these reasons that the inhabitants rejected the government plan, we propose a plan to build elevated houses over the existing road, next to the current settlement (Fig. 12.9), so that the inhabitants could remain in the same area, but with a minimum flood risk, even during the heavy rains. The idea of this plan is mainly considered due to the resistance of the inhabitants against the eviction, while on the other hand, elevating the buildings in the flood area zones is likely to be the best answer to minimise flooding problems (Karyono 2014). The government scheme

Fig. 12.9 The proposed settlement will be built over the existing road

for eviction is considering only the way to rescue people from the floods, but not solving the other problems which have appeared after the evictions.

Based on the interviews we found that there have been some efforts done by the inhabitants to elevate their houses by raising the floor levels. However, as many times as they have done it; the floods have been getting higher and higher all the time. Raising the floor has little effect to avoid the rising water because there are no additional ground surfaces available to absorb the water. To enlarge the ground surface areas for absorbing the river water, the floor areas covering the existing ground surface must be minimised. In other words, the houses must be elevated, building them on stilts, thus leaving the ground surfaces uncovered.

To relocate the inhabitants to a safer settlement, without breaking up their community ties, there is a need to provide new houses first, before the inhabitants are asked to leave their old houses. To build a new settlement, we need an empty space. Since there is no empty space available in this dense area, the new houses will be built on stilts over the nearby existing road (Fig. 12.9). Applying this plan, the inhabitants can remain in their current houses until the new elevated houses are completed. It is expected that the new houses would accommodate properly the people who live in flood-affected shelters.

The construction process should be carried out in such a way that the traffic along the road might not be seriously disturbed during the construction process.

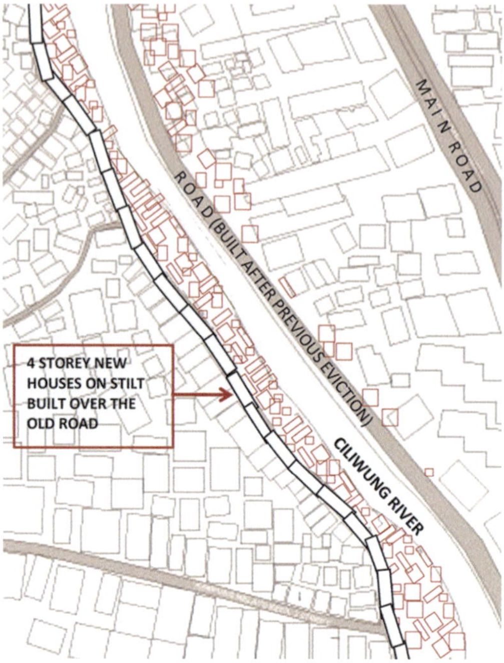

Fig. 12.10 The new 4 storey stilt houses are built over the existing small road, while the road beneath the houses is still used

When the construction process is finished and the new houses are completed, people are asked to leave their old houses and move to the new elevated houses (Fig. 12.10).

Again, during this process, the traffic along the road must be maintained to be safe and undisturbed. When all the inhabitants have moved into the new houses, the old houses are to be demolished. The empty land of the former squatter shelters is then used for a new road, replacing the old one (Fig. 12.11).

At the two ends of the old roads, which are not covered with the new houses, connections to the new road are made to diversify the traffic. The narrow empty land along the river and the new road is then planted to form a green area (Fig. 12.12).

5.2 *Building Design and the Sustainable Features*

Although there is a need to make further analyses regarding the proposed sizes of the new houses, this proposed design has tried to derive the size of the housing units by considering the current houses' sizes collected from the survey. Based on the survey, the sizes of the existing houses were between 6 and 225 m^2. The existing houses of more than 60 m^2 were occupied by more than one family, whereas the proposed house units are considered to be occupied by one family only. In order to simplify the determination of sizes of the new houses, they are designed in a modular system. As the houses are to be built over the road of 5 m width, the span of the

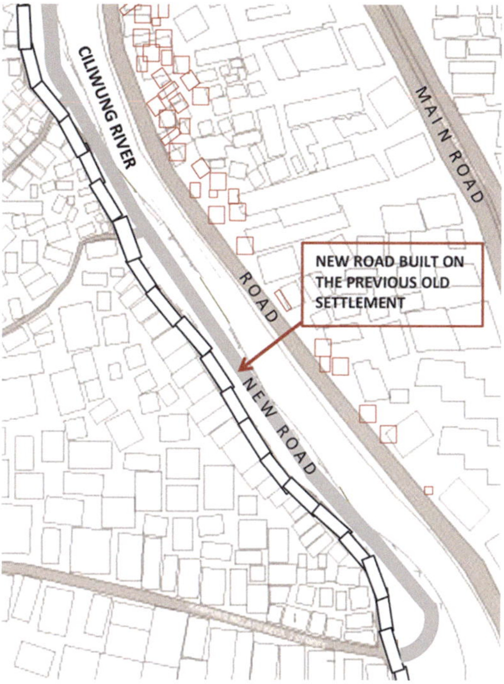

Fig. 12.11 A new road is built on the empty land after the demolition of the old settlement

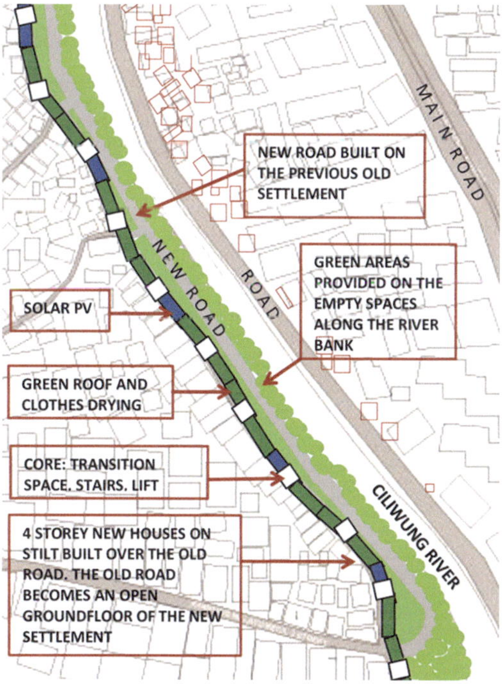

Fig. 12.12 Green areas are formed on the empty spaces between the new houses and the river

house is taken to be 6 m plus a cantilevered access balcony of 1.5 m in one side, whilst the length of the module is 3 m. The size of the smallest house is equal to one modular unit that is 3 × 3 m or 9 m². There will be four house types: A, B, C and D-types. The A-type is 9 m², the B-type is 18 m², the C-type is 36 m² and the D-type is 72 m². Since existing houses of more than 60 m² tend to be occupied by more than one family, it is decided than the largest size of the proposed houses is 72 m². This size is derived from a multiplication of the modular unit of 9 m², and it is provided to accommodate the inhabitants who currently occupy a 60 m² house and larger sizes.

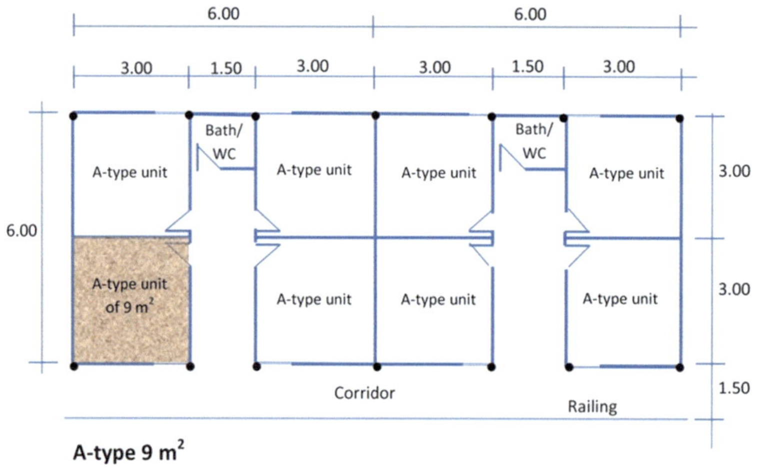

Fig. 12.13 A-type unit of 9 m²

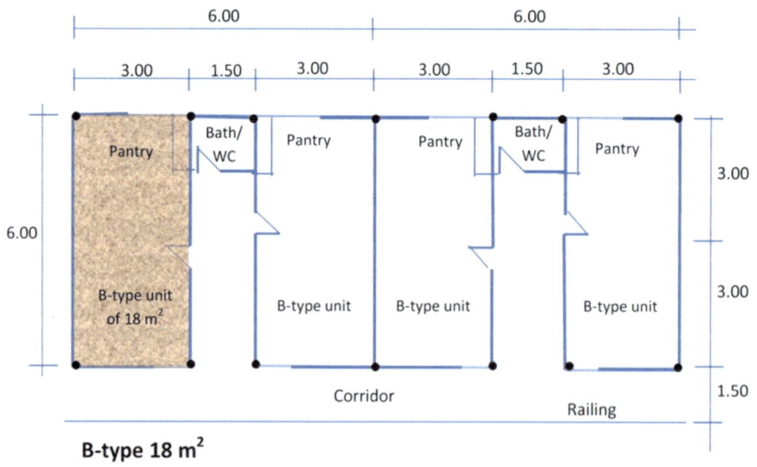

Fig. 12.14 B-type unit of 18 m²

Those with a current house of between 6 and 12 m^2 will be provided for by the A-type, a single module of 9 m^2, the smallest type house unit in the new settlement. Those with current house sizes of 60 m^2 and larger would be provided with eight modules of 9 m^2, forming the D-type of 72 m^2, which is designed in a maisonette (2-floor) form. This maisonette-type consists of four modules on the lower floor and another four on the upper floor. Those currently living in houses between 6 and 60 m^2 would be accommodated in the new houses with the size of between 9 and 72 m^2. Figures 12.13, 12.14, 12.15 and 12.16 show the floor plans of the new house types.

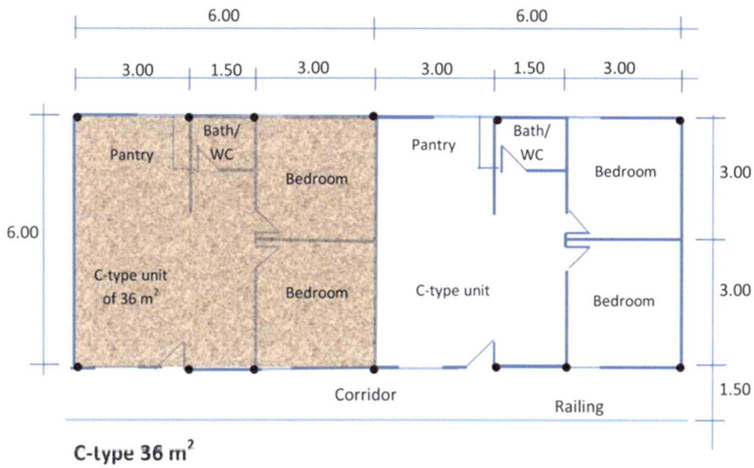

Fig. 12.15 C-type unit of 36 m^2

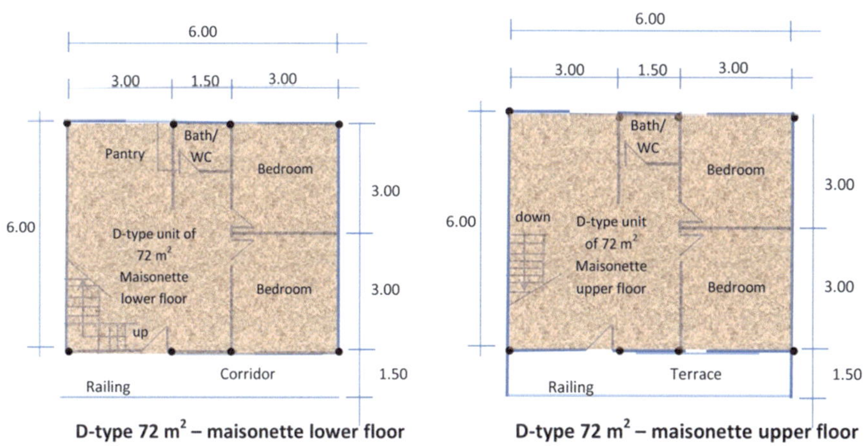

Fig. 12.16 Lower and upper floor of D-type unit of 72 m^2 – maisonette type

There are public showers and WCs provided in the A and B types. Public showers and WCs are commonly used in the current settlement. Therefore, in order to optimize the dwellings of the smaller types of 9 and 18 m^2, the public showers and WCs would be provided to benefit the inhabitants accommodated in these types of dwellings.

The height of the first floor from the ground is no less than 3.5 m, more than adequate considering the height of the water level during the floods is less than 2 m. The dwelling units will be arranged with a single loaded access corridor which will face to the river as there is more space left on this side to build the corridor, stairs and elevators. At every 30 m length along this settlement, there will be a building core, used for structure bracing and also housing an elevator and stairs for vertical circulation and to escape from fire. A foyer of 3 × 6 m is provided in this area, which has a double function as a disconnected space in case of fire and also for a public space and a battery room for the solar PV. The main vertical circulation will be by stairs; however, elevators will be used for elderly people, disabled persons, pregnant women and goods.

Since the dwellings are designed as a modular unit, the arrangement of the individual dwelling units is flexible. A mixed arrangement between A, B, C and D types can be done or, alternatively, a single type of dwelling can be located vertically in a particular section. However, considering the current condition, in which all different sizes of the shelters are mixed together, all types of the new dwellings would be mixed up, without being grouped. Whatever the arrangements of the dwelling types, design of the building should meet the occupants' comfort requirements in the tropical climate, with minimum energy consumption (Karyono 2007).

The ground floor, which was formerly a road, will be paved with concrete blocks to be used for a public area. A number of rooms are going to be provided in this ground floor, such as the community offices (for heads of RTs and RWs), primary health services, community centre, kids' playground, dairy shops, public toilets, and bicycle and motorcycle parking. Water reservoirs, where the water is supplied by the Jakarta water authority, are built beneath the ground floor. The clean water will be pumped up from the reservoirs to the roof tank and distributed to all the dwelling units.

Concrete will be used for the main structure, such as columns and beams. Half of the lower walls will be built with bricks; the rest of the walls are of wood. There will be a flat roof of concrete. Most of the roof will be green, covered by small plants in pots. The roof will have multi functions. Besides being green, some parts of the roof will be used for clothes drying, since there is limited space available for clothes drying in every dwelling unit. Other parts of the green roofs will also be used for solar PVs to supply a proportion of the household electricity demand, along with the supply from the national grid. There will be also water tanks on the slab roofs.

References

BPS-Statistics DKI Jakarta Provincial Office (2006) Climate, Jakarta in figures

Karyono TH (2003) Sekali lagi Dilema Penggusuran (Once again, problems of eviction), Kompas daily newspaper, 25 October 2003

Karyono TH (2004) Hunian Tanpa gusuran: Masalah Hunian Masyarakat Bawah di Kota (Housing without eviction: problems of low-income people live in the city), paper. In: Proceedings of national conference on dwellings for all people. Duta Wacana Christian University, Yogyakarta, 9 October 2004

Karyono TH (2007) Rumah Susun Tropis Hemat Energi (Low energy tropical flats for low income people), paper. In: Proceedings of low energy flats conference. Parahyangan University, Bandung, 8 September 2007

Karyono TH (2012) Frequent cases of flooding reduces Jakarta to third-world city. The Jakarta Post, 10 February 2013

Karyono TH (2014) Floods and raising the city. The Jakarta Post, 5 April 2014

McCarthy P (2003) Urban slums reports: the case of Jakarta, Indonesia, in understanding slums: case studies for the global report on human settlements. World Bank

Widianto B (ed) (2012) Rapid Assessment Daerah Aliran Sungai Ciliwung Di Kelurahan Manggarai & Kelurahan Kampung Melayu. P3KS Press, DKI Jakarta

World Weather Information Service, Indonesia, Jakarta, World Meteorological Organization (2016) Available at http://worldweather.wmo.int/043/c00310.htm. Accessed on 25 June 2016

Part III
Mitigating Building for Climate Change

Chapter 13
Mitigating the Environmental Impacts of the Development of Ecotourism Through Using Refurbished Buildings as Its Related Accommodation Services

Abbas Mahravan

Abstract This article investigates the environmental influence of using refurbished buildings as ecotourism accommodation services in terms of the effect on the host society's environmental sustainability. It places the conservation of natural resources as the main ecological outcome for ecotourism. It uses the ecological footprint of the ecotourism accommodation services as the ecological indicator for evaluation of these services to ensure they are ecologically compatible. The article presents a case study of 'The Otago Central Rail Trail' (OCRT) as an example of ecotourism. The investigation reviews the impacts of the OCRT accommodation services on environmental resources by using ecological footprint. The results of this study suggest that in OCRT, using refurbished buildings as accommodation services contributes to reducing of environmental impacts related to OCRT. This article makes it clear that in OCRT using 12.9 % (21,378 m^2) of bed space (80,356m^2) in refurbished buildings reduces by 14.3 % (7.3 gha) the total 51 gha ecological footprint of accommodation services compared with when all accommodation buildings are assumed to be new buildings. The study described here was undertaken in a temperate climate but the method used would be equally appropriate for use in assessing the impact of facilities in the tropics.

Keywords Accommodation service • Ecological footprint • Refurbished building • Sustainability

A. Mahravan (✉)
Department of Architecture, Razi University, Kermanshah, Iran
e-mail: mahravan_abbas@yahoo.com

T.H. Karyono et al. (eds.), *Sustainable Building and Built Environments to Mitigate Climate Change in the Tropics*, DOI 10.1007/978-3-319-49601-6_13

1 Introduction

This article introduces the idea of using refurbished buildings as ecotourism accommodation services as a way that can contribute to mitigating the environmental impacts of tourism. Conservation of environmental resources is viewed here as the main goal for development of ecotourism, as a more sustainable segment of mass tourism.

Wearing and Neil (2009) recognise ecotourism as one of the segments of the tourism industry which is growing three times faster than other parts. According to Blangy and Mehta (2006), the International Ecotourism Society (TIES) views ecotourism as responsible travel to natural destinations which protects the environmental resources and develops the well-being of the indigenous people. Sustainability is the essential characteristic of ecotourism which basically becomes its main difference with regard to other segments of the tourism industry. More recent definitions have proclaimed some more detailed considerations for ecotourism such that it "must be economically viable, ecologically sensitive, and culturally appropriate" (Wall 1997). According to this definition, if ecotourism and its development cause negative impacts on the environmental resources, visitors, host communities and their economic systems, it cannot be defined as "ecotourism".

Mahravan (2012) introduces ecotourism as a set of purposeful nature based activities that has the ability to change the attitude of its host society to environmental, social and economic problems caused by development of capitalism. This can be viewed as influences exerted by ecotourism on its host destination towards being more sustainable. Mahravan (2012) defined these influences as the cultural footprint of ecotourism and its products and activities on the host destination.

To clarify the relationship between ecotourism and environmental conservation, Buckley (2009) views minimizing impacts on the natural environment as a key criterion for its evaluation. On the other hand, one of the main goals for developing sustainable ecotourism is to protect ecosystems which provide the services essential for human activity. In terms of ecotourism, an ecosystem "in short can be described as supporting life, supplying materials and energy, and absorbing waste products" (Gössling 1999). As Murphy (1983) points out, all planning (including planning for the sustainable development of tourism) needs goals to supply a frame of reference for detailed physical planning and for the constant management decisions that follow. He argues that these goals are abstract and continuous concepts intended to prepare a general direction rather than specific guidelines, but that they must have general support and be feasible to be effective.

Mahravan and Vale (2011) reveal the sustainable development of ecotourism and its goals as socio-ecological phenomena. They propose conservation of environmental resources; awareness of local and indigenous people about their environmental heritage and capitals through a social educational process; and engagement of these educated people in the sustainable development process of ecotourism as the main social-environmental outcomes of ecotourism (see Table 13.1).

Table 13.1 Ecological indicators for evaluation of the environmental impacts of ecotourism development on the host destinations

Ecological outcome	Indicator	Definition
Environmental awareness	1a	The number of local people who participate in the educational process.
Engagement of local and indigenous people in the conservation process.	1b	The number of people who engage in the environ-mental conservation process including volunteer or employed participants.
Attention to carrying capacity and sustainable yield	1e	Ecological footprint of tourism activities, products and services.
Protection of environmental resources and maintenance of biodiversity	1e	Ecological footprint of tourism activities, products and services.
Contribution through use of renewable resources to generate energy used by tourism products and services.	1e	Ecological footprint of tourism activities, products and services.

Mahravan and Vale (2011)

Table 13.1 displays a set of proposed social-ecological indicators for evaluating the environmental impacts of ecotourism development on the host destinations and achieving its anticipated environmental outcomes. Indicators '1a' and '1b' (Table 13.1), which indicate the number of participants in environmental education and the conservation process are also introduced as social cultural indicators. These indicators can also be viewed as complementary indicators to the ecological footprint (EF) to evaluate the success of the social dimension of environmental conservation principles.

2 Ecological Footprint

Ecological footprint (EF) is used here as an ecological indicator to evaluate ecotourism and its products (including refurbished buildings used as ecotourism accommodation services) as being environmentally sustainable. Ecological footprint (EF) is a way of measuring environmental impact. "It assumes that everything needed for living, including all energy, goods and services can be obtained from land, and that any wastes produced can be absorbed by land" (Wackernagel and Rees 1996). Because land productivity varies considerably around the world the normal measure is in global hectares (gha), a global hectare being a hectare of land of average productivity. The ecological footprint of a product or activity produced/consumed or conducted by a defined group of people is measured here through the following equation:

$$EF(gha) = \text{Lifecycle energyuseofagiven group of people}$$
$$\text{(productandactivity produced or conducted)}$$
$$\text{Gigajoules}(GJ) / \text{carrying capacity of the Earth}(GJ / gha)$$

The above equation is based on the area required to produce energy resources. The assumption used is that one gha can provide 100GJ of renewable energy (Wackernagel and Rees, 1996).

3 Case Study: Otago Central Rail Trail

3.1 Background

"Originally the Otago Central Railway that is located in the South Island of New Zealand was constructed to transport gold out from the booming Central Otago goldfields of the late 1800s and into the bustling city of Dunedin" (Graham 2009). In the 40 years it took to complete the railway, over 60 bridges and 3 tunnels were built as part of the railway between Middlemarch and Clyde. According to Burns and Corbet (2007) the railway closed in 1990, at a time when the rural community was in a state of stagnation. Graham (2009) points out that through the early 1990s, towns in Central Otago as in other rural areas in New Zealand lost many of their services and suffered from a reducing population, as banks and post offices closed and many businesses moved away.

However, as explained by Burns and Corbet (2007) some local people viewed the closed railway line as an opportunity and potential attraction for developing the tourism industry in Central Otago. The Department of Conservation (DOC), which had bought the land in 1993, was persuaded to consider the potential for outdoor recreation that the rail land offered. "DOC spent six years and more than $850,000 preparing, upgrading and restoring the closed railway and related historical heritage (including 60 historical bridge and 3 tunnels) and infrastructure, which are now viewed as one of the main attractions for OCRT visitors" (Burns and Corbet 2007).

3.2 OCRT Accommodation Services

3.3 Types of Accommodation Services

As presented in Table 13.2, the total number of 783 OCRT accommodation services is broken down into seven categories; self-contained (SC), bed and breakfast (B&B), backpacker (Bp), motels, homesteads, hotels and camping sites. Table 13.2 shows available bed spaces by types of accommodation service. As presented in Table 13.2,

Table 13.2 OCRT- area of accommodation services by type, number and quality

Category	Q	No	TNA	Average area/bed space (m^2)	Bed space	Area (m^2)
10,355	NB	62	92	29.5	351	10,354.5
4691	RB	30		29.5	159	4690.5
15,280	NB	44	66	40	382	15,280
6360	RB	22		40	159	6360
996	NB	8	12	12	83	996
636	RB	4		12	53	636
10,867	NB	32	33	16.9	643	10,866.7
372	RB	1		16.9	22	371.8
176	NB	1	4	29.4	6	176.4
647	RB	3		29.4	22	646.8
1397	NB	1	19	29.1	48	1396.8
8672	RB	18		29.1	298	8671.8
19,902	NB	576	576	6	3317	19,902
–	RB	–		–	0.00	–
80,350	–	783		–	5543	80,350

Mahravan (2012)
TNA total number of accommodation services types
RB refurbished buildings used as accommodation services
NB new buildings used as accommodation services
Total number of RBs = 78 (10 %)
Total number of NBs= 705(90 %)
The total area of OCRT NB accommodation services = 58,972 m^2
RB accommodation is 21,378 m^2

in 2011, of the total 783 separate accommodation services the majority or 71.1 % (576) are camping sites followed by 11.8 % (92) SC and 8.5 % (66) B&B. There are small numbers of other types of OCRT accommodation services including Bp, motels, homesteads, and hotels in the range of 1.5–4.2 % of the total (Table 13.2).

3.3.1 Quality of OCRT Accommodation Services

As shown in Table 13.2 OCRT accommodation services are classified into the two types of new (NB) and refurbished (RB) buildings. In OCRT, RB accommodation services cover a range of historical and restored buildings with different original functions, but currently all of them are used for visitor accommodation. Table 13.2 shows the number and percentage of NB and RB buildings used as accommodation services in the OCRT and demonstrates that of 781 OCRT accommodation services, only 10 % (78) are RB and 90 % (705) are NB.

Table 13.2 indicates that of the 10 % (78) RB accommodation services, the majority or 3.8 % of the total (30) are SC, followed by 2.8 % (22) B&B, and 2.3 % (18) hotel. Other RB types of accommodation including Bp, motel, homestead and camping only make up 0–0.5 % of the total. Moreover as shown in Table 13.2, of 90 % of total (705) NB accommodation services, the majority or 71.1 % (557) are camping sites (powered and non-powered) followed by 7.9 % (62) SC, 5.6 % (44) B&B and 4.1 % (32) motel. Other NB types of accommodation services including B&B, Bp and hotel only make up 0.1–1 % of total (Table 13.2).

3.3.2 Accommodation Services – Area

Table 13.2 shows the average area, number of bed spaces and total area of each type of OCRT accommodation service and whether they are new (NB) or refurbished (RB). In this Table the area of each type of accommodation services is calculated as its average area per bed space multiplied by the number of bed spaces. Comparing the areas of types of NB OCRT accommodation services, camping sites comprise the majority or 19,902 m^2 followed by B&B (15,280 m^2), motel (10,867 m^2) and SC (10,355 m^2) (Table 13.2). Other types of NB accommodation services including Bp, hotel and homestead contain areas in a range of 176–1397 m^2 (Table 13.2). For refurbished buildings, RB hotels contain the largest RB area of 8671 m^2 followed by B&B (6360 m^2) and SC (4691 m^2). The areas of other refurbished accommodation services including Bp, hotel and homestead are between 372 and 647 m^2 (Table 13.2). The total area of OCRT NB accommodation services is 58,972 m^2 and RB accommodation is 21,378 m^2, making a total of 80,350 m^2 (Table 13.2).

3.3.3 OCRT Accommodation Services: Materials Used

The main materials used for constructing each type of OCRT accommodation services (except camping sites) are summarised in Table 13.3 where the information has been categorised into materials used by type, number and capacity of OCRT accommodation services in 2011.

The types of materials used are classified into the eight types of timber, timber and masonry, mud brick (mixed with other materials), brick (mixed with other materials), stone, concrete, containers, and caravans (Table 13.3). As shown in Table 13.3, of the total 2226 building based OCRT accommodation bed spaces the majority 49.1 % (1094) of bed spaces are in timber buildings followed by 28.3 % (628) timber and masonry, 8 % (178) stone, 5.7 % (126) brick (mixed with other materials) and 5 % (113) concrete. Other materials used including mud brick (mixed with other materials), containers and caravans make up 0.2–3.5 % of total (Table 13.3).

The SC accommodation services comprise 388 (76 %) bed spaces using timber as the main construction material followed by 55 (10.8 %) timber and masonry and 33 (6.5 %) mud brick mixed with other materials. Bed spaces that are in buildings of other materials including stone, brick, concrete and caravans/containers fall in the range of 4 (1.2 %)–20 (3.9 %) out of the total (Table 13.3).

Of 541 B&B bed spaces the majority or 225 use timber as main construction material, followed by 169 timber and masonry and 50 brick (mixed with other materials). The numbers of B&B bed spaces of other materials including mud brick (mixed with other materials) and stone are between 26 and 37 of the total bed

Table 13.3 OCRT – accommodation services – materials used by types of accommodation (2011)

Material (s)	Total bed spaces	%	Types of accommodation services											
			SC		B&B		Bp		Motel		Homestead		Hotel	
			No	C	No	C	No	C	No	C	No	C	No	C
Timber	1094	49.1	72	388	29	225	6	54	26	357	1	6	4	64
Timber and masonry	628	28.3	8	55	14	169	3	54	3	163	–	–	7	187
Mud brick (mixed with other materials)	79	3.5	5	33	4	26	1	6	–	–	1	6	1	8
Brick[a] (mixed with other materials)	126	5.7	1	6	7	50	1	18	1	44	–	–	1	8
Stone[b]	178	8	4	20	8	37	1	4	1	22	2	16	6	79
Concrete	113	5	–	–	4	34	–	–	2	79	–	–	–	–
Container	4	0.2	1	4	–	–	–	–	–	–	–	–	–	–
Caravan	4	0.2	1	4	–	–	–	–	–	–	–	–	–	–
Total	2226	100	92	510	66	541	12	136	33	665	4	28	19	346

Mahravan (2012)
No number of accommodation services
C capacity (number of bed spaces)
[a]Brick includes: red brick, clay brick and brick
[b]Stone includes: schist stone, stone and rock

spaces. As shown in Table 13.3, 108 (79.4 %) of 136 Bp bed spaces are constructed from timber and timber (mixed with other materials). The other 28 Bp bed spaces used materials including mud brick (mixed with other material, brick (mixed with other materials), stone and concrete, forming between 4 (3 %) and 18 (13.20 %) of total in this category

The majority of 357 (53.7 %) motel bed spaces were built in timber followed by 163 (24.5 %) timber and masonry, 79 (11.9 %) concrete, 44 (6.6 %) brick (mixed with other materials), and 22 (3.3 %) stone (Table 13.3). In contrast, of 28 homestead bed spaces, the majority of 16 (57.2 %) were in stone followed by equal numbers of timber, and mud brick (mixed with other materials), at 6 (21.4 %) each. OCRT hotels comprise 346 bed spaces of which 187 (54 %) are built of timber and masonry, 79 (22.8 %) stone, and 64 (18.5 %) timber. Other hotel bed spaces are equal in number between mud brick and brick mixed with other materials at 8 (2.3 %) each.

3.3.4 All Types of OCRT Accommodation Services: Energy Sources and Facilities (2011)

Figure 13.1 indicates the energy resources used and facilities provided in all types of OCRT accommodation services. It shows of 5496 OCRT bed spaces, 3233 (58.80 %) NB and 621 (11.30 %) RB use electricity for heating, also 121 (2.2 %) NB and 258 (4.7 %) RB use logs for heating, 324 (5.9 %) NB and 88 (1.6 %) RB use gas, and 77 (1.4 %) NB and 16 (0.3 %) RB use a solar system for heating water. When it comes to equipment, Fig. 13.1 shows that of 5496 OCRT bed spaces, 401 (7.3 %) RB and 687 (12.5 %) NB bed spaces have TV (shared and private) and only 121 (2.2 %) RB have access to the Internet compared with 3985 (72.5 %) NB bed spaces. in addition, 278 (5.60 %) RB and 539 (9.80 %) NB bed spaces use shared and private outdoor sitting areas respectively (including verandas, balconies and open air sitting spaces).

3.3.5 OCRT – Accommodation Services – Occupancy Rate

Table 13.4 presents calculated occupancy rates of Central Otago accommodation services in 2011. Table 13.4 shows that the occupancy rate of camping in OCRT is 24 % followed by motel (21.1 %), backpackers (14 %), B&B (13.9 %), SC (13.1 %), hotel (10.9 %) and homestead (3 %).

In addition Table 13.4 indicates that the number of visitor nights of camping in OCRT is 10,185 followed by motel (8954), Bp (5941), B&B (5899), SC (5559), hotel (4626) and homestead (1273).

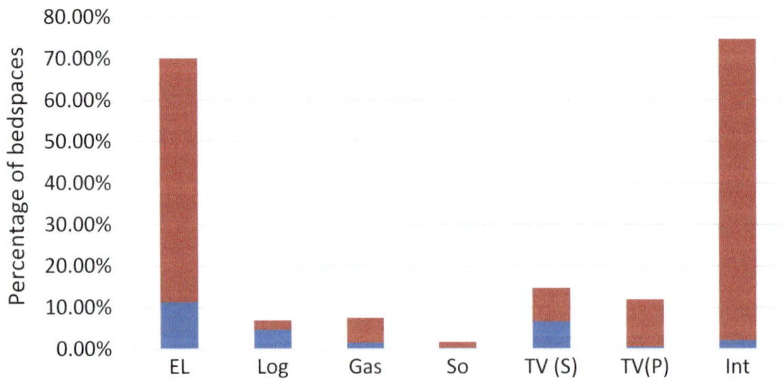

	EL	Log	Gas	So	TV (S)	TV(P)	Int
NB	58.80%	2.20%	5.90%	1.40%	8%	11.30%	72.50%
RB	11.30%	4.70%	1.60%	0.30%	6.70%	0.60%	2.20%

Fig. 13.1 OCRT- all types of accommodation services- energy sources and facilities (2011) (Mahravan 2012)

Table 13.4 OCRT- accommodation services-rates of occupancy (2011)

Type of accommodation service	% of visitor nights	Visitor nights
SC	13.1	5559
B&B	13.9	5899
Bp	14	5941
Motel	21.1	8954
Homestead	3	1273
Hotel	10.9	4626
Camping	24	10,185
Total	100	42,437

Mahravan (2012)

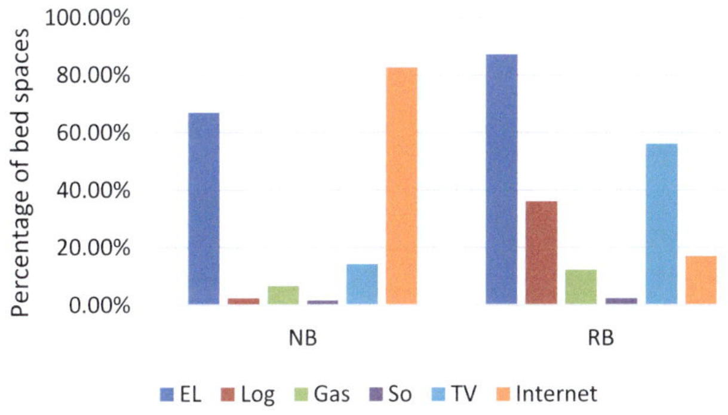

Fig. 13.2 Comparison between OCRT – NB and RB accommodation bed spaces in terms of energy sources, facilities and open air spaces

3.3.6 OCRT Accommodation Services: Comparison Between NB and RB Bed Spaces in Terms of Energy Sources, Facilities and Open Air Spaces (2011)

Figure 13.2 compares OCRT NB and RB bed spaces for use of energy resources and facilities. As demonstrated, 66.5 % of NB and 87 % of RB bed spaces use electricity for heating. However, only 2.5 % of NB have facilities for log burning compared to 36.20 % of RB bed spaces. Use of gas is more even with 6.70 % of NB and 12.30 % of RB bed spaces using it for heating. Fig. 13.2 also indicates that 1.60 % of NB and 2.40 % of RB bed spaces use a solar system for heating water, showing that both types of buildings can accommodate more sustainable technologies.

Figure 13.2 also shows that 14.22 % of NB and 56 % of RB bed spaces provide a TV, but it should be remembered that the NB bed spaces include 3317 camp bed spaces that do not provide TV. In terms of using the Internet, 82.50 % of NB and only 17 % of RB bed spaces have access to the web (Fig. 13.2). There are some clear differences in what visitors can expect to find in the different types of accommodation.

4 The EF of OCRT-Accommodation Services

The Ecological footprint (EF) of OCRT accommodation services is explored through using two scenarios. In the first scenario, all buildings are assumed as new constructed buildings (NB) and in the second scenario, as in the present circumstance of OCRT accommodation services the buildings include both NB and RB accommodations. Comparison between the EFs arising from the first and second scenarios serve to indicate the influence exerted on the EF of OCRT- accommodation services by using refurbished buildings.

4.1 *First Scenario: All Buildings Are Assumed to Be New Buildings*

As mentioned in Sect. 3, calculation of the EF of OCRT accommodation services in this study involves four factors; embodied energy use; operating energy use; lifecycle energy use and carrying capacity of the earth. This section presents the measurement of embodied energy use and operational energy use related to OCRT accommodation services which are then used to calculate lifecycle energy use and EF of the services.

4.1.1 OCRT – Accommodation Services – Embodied Energy

Table 13.5 presents integrated embodied energy of wall, roof, floor and foundation used in each type of OCRT accommodation construction. As shown in Table 13.5, in comparison with other types of OCRT construction, brick mixed with other materials (old technology) construction has the largest embodied energy of 5.12 GJ/m^2 and timber structure has the lowest. The embodied energy of other types of OCRT-constructions are in a range of 1.4–3.15 GJ/m^2. Embodied energy of OCRT accommodation services by types of construction (Table 13.5) is calculated through the following equation:

$$\begin{aligned}&\text{Embodied energy } (GJ\,/\,year) \text{ of each type of construction}\\&= \text{Total area } (m^2) \text{ of the type of construction}\\&\times \text{ Embodied energy use (over 50 – years life cycle in } GJ\,/\,m^2)\\&\quad \text{of the type of construction } (\textit{see} \text{ Table 13.5}) \div 50.\end{aligned}$$

Table 13.6 shows the total embodied energy of each type of OCRT accommodation services. As indicated in this Table, of 2095.8 GJ/year of embodied energy for OCRT accommodation services, the majority of 840.2 GJ/year is the embodied

Table 13.5 Embodied energy of OCRT – accommodation services by types of construction (2011)

Material	EE[a] over a 50-year life cycle (GJ/m^2)	SC-		B&B		Bp		Motel		Homestead		Hotel	
		TA[b] (m^2)	EE (GJ/y^3)	TA (m^2)	EE (GJ/y)	TA (m^2)	EE (GJ/y)	TA (m^2)	EE (GJ/y)	TA (m^2)	EE (GJ/y)	TA (m^2)	EE (GJ/y)
Timber	1.1	11,466	252.2	9000	198	648	14.3	6033	132.7	176	3.9	1862	41
Timber and masonry	2.04	1622.5	66.2	6760	275.8	648	26.4	2755	112.4	–	–	5442	222
Mud brick (mixed with other materials)	1.4	973.5	27.3	1040	29.12	72	2	–	–	176	4.9	233	6.5
Brick mixed with other materials (old technology)	5.12	177	18.1	2000	204.8	216	22.1	744	76.2	–	–	233	23.9
Stone (local)	1.58	590	18.6	1480	46.8	48	1.5	372	11.8	470	14.9	2299	72.6
Concrete	3.15	–	–	1360	85.7	–	–	1335	84.1	–	–	–	–
Total			382.4		840.2		66.3		417.2		23.7		366

G/Y gigajoule per year
[a]*EE* embodied energy use
[b]*TA* total area

Table 13.6 OCRT-Embodied energy used of accommodation services (2011)

Types of accommodation	Total Embodied energy (GJ/year)
SC	382.4
B&B	840.2
Bp	66.3
Motel	417.2
Homestead	23.7
Hotel	366
Camping (powered sites)	NA
Total	2095.8

Table 13.7 Mean energy intensities for various accommodation categories

Category	Energy use (GJ/m²/year)	Occupied area (m²)	Total operating energy use (GJ/year)
SC	0.3	1971	591.3
B&B	0.3	3008	902.4
Bp	0.61	228	139
Motel	0.25	2371	593.7
Homestead	0.25	25	6.3
Hotel	0.57	1097	625.2
Camping (powered sites)	0.025	–	156.2
Total			3014.1

Becken et al. (2001)

energy of B&Bs followed by motels (417.2), SC (382.4), hotels (366), Bp (66.3) and homesteads (23.7).

4.1.2 OCRT – Accommodation Services – Operating Energy

As shown in Fig. 13.2 OCRT accommodation services (NB and RB) use conventional sources of energy and facilities and similar operating energy for heating and lighting of their bed spaces. Base on this fact, operating energy figures for each type of New Zealand accommodation services as proposed by Becken et al. (2001) are used here. As indicated in Table 13.7, in New Zealand, the Bp category has the largest energy demand of 0.61 (GJ/m²/year) followed by hotel (0.57 GJ/m²/year), SC and B&B (each 0.3 GJ/m²/year), motel, homestead and campgrounds (each 0.25 GJ/m²/year).

In Table 13.7 operating energy use (GJ/year) of each type of accommodation services is equivalent to occupied area of the type of accommodation × operation energy use (GJ/m²/year) of the type of construction. In this equation the occupied area of each type of construction is equivalent to % of occupancy rate of the type of accommodation × total area (m^2) of the type of construction (see Table 13.8). Likewise, in above equation, total area (m^2) of the type of construction' is calculated as following: Total area (m^2) of the type of construction = Capacity of the type of accommodation services (see Table 13.8) × average area/ bed of the type of accommodation services (see Table 13.8).

4.1.3 OCRT – Accommodation Services – Life Cycle Energy Use

Life cycle energy use of each type of OCRT accommodation is calculated as the sum of embodied energy and operation energy use of that type of accommodation (Table 13.9). Table 13.9 indicates that of 5109.9 (GJ/year) total OCRT lifecycle energy use, B&B uses the majority of 1742 (GJ/year) followed by motel (1010), hotel (991.2), SC (973.7), Bp (205.3), camping (156.2) and homestead (30).

Table 13.8 OCRT- accommodation services- total area and occupied area (2011)

Type	No.	C	Area per bed m^2	Total area	Occupancy (%)	Occupied area m^2 (2011)
SC	92	510	29.5	15,045	13.1	1971
B&B	66	541	40	21,640	13.9	3008
Bp	12	136	12	1632	14	228
Motel	33	665	16.9	11,238.5	21.1	2371
Homestead	4	28	29.4	823.2	3	25
Hotel	19	346	29.1	10,068.6	10.9	1097
Camping	557	3317	6	19,909	24	–
Total	783	5543		80,350	100	

Table 13.9 OCRT – EF of accommodation services (first scenario)

Types of accommodation	Embodied energy	Operating energy	Life cycle energy use (GJ/year)	EF (gha/year)
SC	382.4	591.3	973.7	9.7
B&B	840.2	902.4	1742.6	17.4
Bp	66.3	139	205.3	2
Motel	417.2	593.7	1010.9	10.1
Homestead	23.7	6.3	30	0.3
Hotel	366	625.2	991.2	9.9
Camping (powered sites)	N/A	156.2	156.2	1.6
Total	2095.8	3014.1	5109.9	51

4.1.4 The Total EF of OCRT – Accommodation Services (First Scenario)

The total EF of 788 OCRT-accommodation services (based on the assumption in the first scenario that all are NB) is 51 gha/year in 2011(Table 13.9). As demonstrated in Table 13.9, in comparison with other types of OCRT accommodation services, B&Bs have the biggest EF (17.4 gha) and homestead accommodations have the smallest EF (0.3 gha). Other types of OCRT accommodation services have EFs in a range of 1.6–10.1 gha.

4.2 The EF of OCRT – Accommodation Services (Second Scenario: Reducing Embodied Energy Through the Use of Refurbished Buildings)

In the second scenario the embodied energies of RB buildings are subtracted from the whole lifecycle energy use of OCRT accommodation services as the buildings already exist. Consequently, the lifecycle energy use of OCRT accommodation services in the second scenario includes only the embodied energy of NB buildings plus total operating energy of both RB and NB buildings. As RB accommodation services use similar energy resources and facilities as NB buildings, their operation energy uses are assumed to be the same as the NBs' operation energy and form part of the whole life cycle energy use of OCRT accommodation services in this scenario (Table 13.10).

Table 13.10 demonstrates that in the second scenario total lifecycle energy use of OCRT accommodation services is equal to 4369.2 GJ/year. Moreover the EF of OCRT accommodation services in this scenario is equivalent to 4369.2 GJ/year ÷ 100 (GJ/ha) = 43.7 gha (Table 13.10).

Table 13.10 OCRT – accommodation services – life cycle energy of new and refurbished buildings (second scenario)

Type	Total embodied energy (GJ/year)	NB (%)	RB (%)	Embodied energy of NB (GJ/year)	Total operating energy (GJ/ year)	Life-cycle energy (GJ/year)	EF (gha/year)
SC	382.4	68.8	31.2	263	591.3	854.3	8.54
B&B	840.2	70.6	29.4	593.2	902.4	1495.6	14.95
Bp	66.3	61	39	40.4	139	179.4	1.79
Motel	417.2	96.7	3.3	403.4	593.7	997.1	9.97
Homestead	23.7	21.4	78.6	5	6.3	11.3	0.11
Hotel	366	13.9	86.1	50.1	625.2	675.3	6.75
Camping	N/A	100	0.00	N/A	156.2	156.2	1.56
Total	2095.8	–	–	1355.1	3014.1	4369.2	43.7

Table 13.11 Reduced EF of OCRT accommodation services influenced by the use of RB as tourism accommodation

EF1 (all buildings are assumed to be new buildings)	Present EF
51 gha/year	43.7 gha/year

Reduced EF of OCRT influenced by the use of RB (12.9 % of total bed spaces) as a part of accommodation services = 7.3 gha/year = 14.3 % of EF1

Table 13.11 compares the EF of OCRT accommodation services from the two scenarios. As shown in Table 13.11, influenced by using RB buildings as accommodation services, the EF of OCRT accommodation is decreased by 14.3 % (7.3 gha) from 51 gha to 43.7 gha. The main reason for reducing the EF of OCRT accommodation services in the second scenario compared with the first is the reduced embodied energy of RB buildings.

5 Conclusion

This article introduces the idea of using refurbished buildings as ecotourism accommodation services as a strategy that contributes to mitigating ecotourism's environmental impact on environmental resources. Through presenting a case study of the OCRT as an example of ecotourism, this article investigates the reduced portion of the ecological footprint of OCRT influenced by using refurbished buildings as its related accommodation services. The results of this study indicate that in the OCRT, using 21,378 m^2 of refurbished buildings to provide bed spaces (out of a total of 80,356 m^2) contributes to reducing its total 51 gha ecological footprint by14.3 % (7.3 gha) compared with when all accommodation buildings are assumed to be new buildings.

References

Becken S, Frampton C, Simmons D (2001) Energy consumption pattern in the accommodation sector-the New Zealand case. J Ecol Econ 39:371–386

Blangy S, Mehta H (2006) Ecotourism and ecological restoration. J Nat Conserv 14:233–236

Buckley R (2009) Ecotourism, principles & practices. CABI, Wallingford

Burns G, Corbet R (2007). New Zealand tourism strategy 2015. New Zealand Ministry of Tourism. Available at, http://www.nztourismstrategy.com. Accessed on 13 Nov 2010

Gössling S (1999) Ecotourism: a means to safeguard biodiversity and ecosystem functions? J Ecol Econ 29:303–320

Graham OJ (2009) The Otago central rail trail: preservation of heritage sites through development for visitor use. In: 3rd Australasian engineering heritage conference. Available at, http://www.ipenz.org.nz. Accessed on 21 Mar 2011

Mahravan A (2012) Eco-tourism and its architecture: a methodological framework for assessing progress towards sustainable development. Published PhD thesis, Victoria University of Wellington

Mahravan A, Vale B (2011) Cultural framework and related indicators for evaluation of sustainability through development of tourism. Presented and published in Shanghai international conference on social science (SICSS 2011), Shanghai

Murphy PE (1983) Tourism as a community industry. J Tour Manag 4(3):180–193

Wackernagel M, Rees W (1996) Our ecological footprint: reducing human impact on the Earth. New Society Publishers, Gabriola Island

Wall G (1997) Is ecotourism sustainable? J Environ Manag 21:483–491

Wearing S, Neil J (2009) Ecotourism, impacts, potentials and possibility. Butterworth-Heinemann, Oxford

Chapter 14
Development of JKR/BSEEP Technical Passive Design Guideline for the Malaysian Building

C.K. Tang

Abstract The Building Sector Energy Efficiency Project (BSEEP) in Malaysia, developed by the JKR (Public Works Department) and UNDP (United Nations Development Programme) was aimed at reducing the growth rate of GHG emissions of the building industry in Malaysia. This technical guideline is required to be made relevant to the building industry, addressing immediate concerns from building design professionals, providing adequate technical knowhow and providing an option for financial saving estimates, while at the same time keeping the guidelines simple to read, understand and use. A stakeholder dialogue session was made at the start of the project and it provided the basis for the issues that should be addressed by the proposed Passive Design Guideline. In total ten chapters were developed.

Keywords Passive design • Energy efficiency

1 Introduction

This Energy Efficiency Passive Design Technical Guideline was written specifically for the Malaysian climate zone. It is an attempt to provide a simple and yet useful guideline to practicing building designers in Malaysia for design decisions to be made quickly for the promotion of energy efficiency in building. An industry dialogue was carried out on the 13th June, 2012 to gain an understanding of the status of current energy efficiency design practices and to identify the information that is sought by the industry, for energy efficiency to be practiced in new building development. Eighteen industry leaders from both the private and public sectors attended the dialogue session.

A "wish list" of information requested by the industry was created from the industry dialogue session. From the "wish list", sets of building energy simulation studies (using climate data for Malaysia) were developed and conducted to provide

C.K. Tang (✉)
Veritas Environment Sdn Bhd, Kuala Lumpur, Malaysia
e-mail: ck.tang@veritas.com.my

T.H. Karyono et al. (eds.), *Sustainable Building and Built Environments to Mitigate Climate Change in the Tropics*, DOI 10.1007/978-3-319-49601-6_14

the foundations for the recommendations made in the finished Guideline. Due to limited time to produce the Guideline, only one building model was developed from which to derive all the estimated energy reductions. Different building sizes, shapes and models will yield slightly different results. However, the results and recommendations from this document are deemed accurate enough as a general guide to make an informed building design decision quickly in the interest of energy efficiency in building. The default building model used for the studies was based on a hypothetical office building scenario of 17 floors with a floor utilization efficiency of 78 % and a gross floor area (GFA) of 38,403 m^2. The Guideline is organised into ten chapters.

2 Chapter 1 – Fundamentals

The first chapter addresses the key fundamentals in energy efficiency in building. These are:

1. Energy efficiency in building can only be achieved by addressing the whole building holistically. i.e. it is impossible to have more than 50 % energy reduction in building by addressing one or two features in a building, however if 50 or more features are addressed, it is possible to have an energy reduction of 50 % or more.
2. The 1st Law of Thermodynamics was explained in this initial chapter within a building scenario to establish a sense of energy flow in building for practising architects and engineers to understand that energy cannot just disappear from a building.
3. Various types of air properties, such as dry, wet and dew point air temperature, differences between relative humidity and moisture content in air were also explained in this chapter.
4. Various types of heat transfer mechanisms of conduction, convection, radiation and phase change were also addressed in this chapter.
5. The important thermal comfort parameters such as operative temperature, Fanger's comfort model and the adaptive thermal comfort model were also presented in this chapter.

These fundamentals were important to be established within the first chapter for this UNDP-JKR technical guideline because all the other chapters will reference these basic fundamentals of building science.

3 Chapter 2 – Weather Data

A clear understanding of Malaysia's weather data enables the designer to design a building that responds to the climate instead of against it. The climate in Malaysia is fairly consistent daily over the entire year; therefore it is useful to get an overview of an average day of hourly performance and the maximum and minimum hourly weather data for a full year.

This chapter provided information on dry bulb temperature, wet bulb temperature, relative humidity, humidity ratio (moisture content), dew point temperature, global radiation, direct radiation, diffuse radiation, cloud cover and wind speed & direction, effective sky temperature and ground temperature of the Malaysian climate. The energy efficiency opportunities and risks offered by the climate is are also presented in this chapter. Many charts were provided in the Guideline for ease of understanding the data and a table of raw data is provided for users who wish to make further use of these data for more in-depth analysis on their own.

4 Chapter 3 – Building Form, Orientation and Core Location

Every architect was taught originally about the importance of building orientation in terms of energy efficiency. However, the evaluation of building orientation cannot be made without including the building form and core location because these items have a significant influence on the overall energy efficiency of a building.

The selection of building form often dictates where the core location can be placed, while the core location dictates whether a mechanical system is used to ventilate spaces such as toilets. Moreover, it is also possible to place the core location as a buffer zone to reduce the impact of solar radiation on the air-conditioned space of a building. The selection of core location would then dictate the main orientation of the building in terms of exposure of air-conditioned spaces to direct solar heat gain. In addition to solar heat gain, exposing the façade to a particular orientation may also increase infiltration rate due to the exposure to higher wind velocities.

Chapter 3 takes these factors into consideration and provides simple guidance on the impact of building form, core location and orientation to allow designers to understand the influence of their decision at an early stage of building design. It was shown in this chapter that the worst building form has an energy consumption 7 % higher than the best building form in terms of energy efficiency (Fig. 14.1).

5 Chapter 4 – Daylight Harvesting

Daylight harvesting in buildings may seem to be a very simple task but it is a subject that need to be mastered well in order to be implemented successfully in this climate zone. There are many design issues that need to be addressed adequately. An attempt is made in this chapter to address all the key issues related to daylight harvesting within the context of the Malaysian climate.

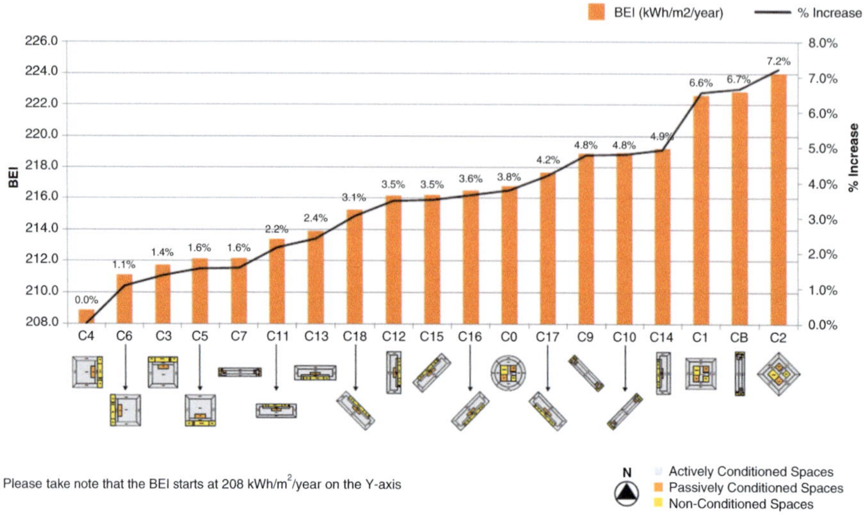

Fig. 14.1 Building energy index of different building form, orientation and core location based on malaysian climate zone

The tropical Malaysian climate is ideal for daylight harvesting to be practiced in office buildings because daylight is consistently available daily from the hours of 8 am to 6 pm. Since Malaysia is located near the equator, there is hardly any variability of season that changes the daily availability of daylight.

An adjustment of Daylight Factor to Malaysia was made in this Guideline, showing that a daylight factor of 0.5 % will yield for more than 90 % of the hours from 9 am to 5 pm a light level of more than 100 lux from diffuse radiation alone (Figs. 14.2 and 14.3).

For a good quality daylight harvesting strategy to be implemented in a tropical climate, the following design principles have to be addressed:

- solar heat gain minimisation,
- glare protection,
- deep daylight penetration and
- uniform daylight distribution

Each of the design principles above is discussed in detail in the Guideline and examples are provided. In addition, various charts were also provided in this chapter for quick estimation of depth of daylight penetration from a basic implementation of daylight harvesting system using horizontal venetian blinds.

Fig. 14.2 Daylight factor adjustment to Malaysian climate, graph shows the percentage hours in a year exceeding 100 lux level from the daytime hours of 8 am to 6 pm and 9 am to 5 pm at various daylight factors

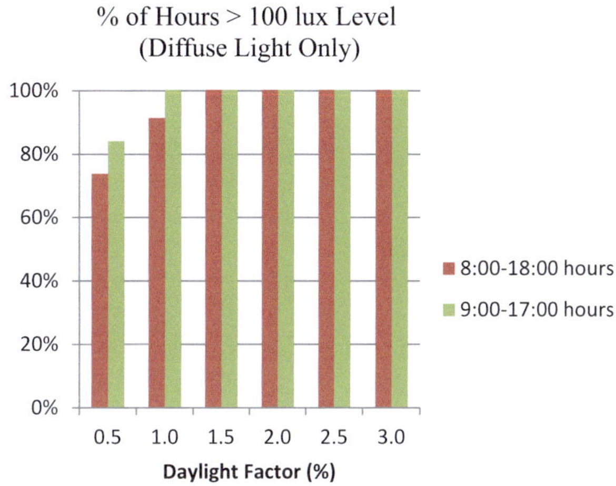

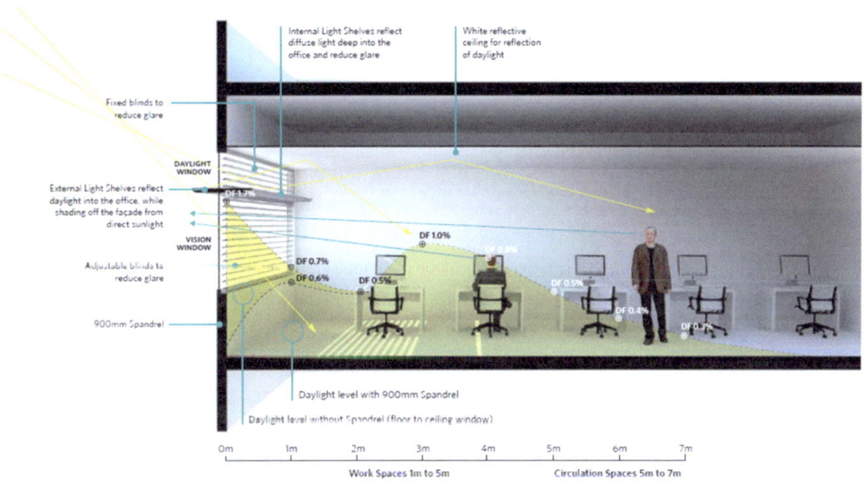

Fig. 14.3 Classic daylight harvesting strategy from Façade showing a split window design for daylight and vision window, incorporating light shelves and glare protection blinds

6 Chapter 5 – Glazing Properties

The selection of correct glazing properties can help to increase energy efficiency in the building and yet provide the 'desired look' of the building. Right glazing selection can reduce the cooling load in the building, it provides better comfort while allowing smaller air-conditioning system to be used, saving investment cost. In addition, selection of right glazing also influences the potential of daylight harvesting in the building that will reduce electric light usage which saves a significant amount of energy in the building, as well as providing improved thermal and visual comfort environment to the building occupants (Boyce et al. 2003).

Many architects today understand the need to have low solar heat gain coefficient (SHGC) to reduce heat gain in the building and high visible light transmission (VLT) to promote daylight harvesting. However, not many understand the relationship between SHGC and VLT due to different glazing technology and mechanism used. This, among many issues with regards to glazing properties selection, is addressed in this chapter.

The ratio of VLT over SHGC provides a term called light to solar gain ratio (LSG). Single glazing without low-e properties has LSG values of 0.5–1.0 depending on the colour of the glazing. Single glazing with low-e properties has typical LSG values of 1.05–1.25. High performance, double glazing with low-e properties has typical LSG values of 1.6–2.0. It is recommended not to use double glazing with LSG values less than 1.5, because the marginal cost of having double glazing with LSG value above 1.5 is relatively low once double glazing is selected to be used in the first place.

7 Chapter 6 – External and Internal Shading

The use of external shades has been well promoted by many architectural books as an essential solution to energy efficiency and thermal comfort in the tropical climate. Meanwhile, improvement in glazing technologies (Chapter 5) has enabled buildings to be built today without the use of external shading devices while complying with respective countries' energy codes. In addition, there exist internal shading devices in the market that claim to reduce solar heat gain in building by 80 % or more. The question of whether it is beneficial to combine all these technologies together and how should this combination be made to optimize the efficiency for the building is addressed in this chapter.

The total SHGC of any fenestration system can be estimated using the following equation (ASHARE 2009):

$$SHGC_{total} = SHGC_{ext} \times SHGC_{glz} \times SHGC_{\text{int}}.$$

Where,

$SHGC_{total}$: Solar Heat Gain Coefficient of the entire fenestration unit.
$SHGC_{ext}$: Solar Heat Gain Coefficient of external shading devices (1, if no external shading device is used)
$SHGC_{glz}$: Solar Heat Gain Coefficient of the glazing.
$SHGC_{int}$: Solar Heat Gain Coefficient of internal shading devices (1, if no internal shading devices is used)

Table 14.1 provides an indication of the potential SHGC total based on various possible combination of external shades, glazing properties and internal shades. A

Table 14.1 SHGC Total computed from various combinations

Descriptions	SHGC ext shades	SHGC glazing	SHGC int shades*	Computed SHGC total
Poorly designed façade	1.00	0.87	1.00	**0.87**
Only 1 item done well	1.00	0.30	1.00	**0.30**
Only 1 item done well	1.00	0.87	0.30	**0.26**
Two (2) items done moderately well	0.70	0.50	1.00	**0.35**
All 3 items done moderately well	0.70	0.50	0.70	**0.25**
All 3 items done well	0.50	0.30	0.50	**0.08**

* ASHRAE fundamentals, F15, Fenestration, Table 13

SHGC reduction of approximately 90 % was shown to be possible from a base case of clear glazing to a case where a good external shade, glazing properties and internal shade are used.

8 Wall Insulation

Energy simulation was conducted to derive an approximate estimate of energy and peak load reductions as a result of using an insulated wall. These estimates are provided as a guide for quick design checks by architects, engineers and building owners to estimate the cost saving of implementing energy efficiency features. The simulation study model was based on Chapter 4, Case 1 of a Square Building, without any external shades. The Case 1 study model is almost a replica of the building model used by the Deringer Group, Berkeley, California, to derive the OTTV (Overall Thermal Transmission Value) constants for all type of buildings for Malaysia, Singapore, Indonesia, the Phillipines and Thailand, under the ASEAN-USAID program in the 1980s – 1992 (Deringer and Busch 1992). The only difference is that the current model building size, floor efficiency and operating assumptions were updated to current standard of practice. The OTTV formulation is still in use today and deemed to be accurate for all types of buildings use, shape, size and hours of operation in these countries.

The energy saved due to the use of insulated walls is dependent on the internal heat load during night time (night time parasitic load). Three scenarios were created where the small power (equipment) night time parasitic load is set to 50 % (high night time parasitic load), 35 % (mid night time parasitic load) and 10 % (low night time parasitic load). A "Wall Simplified Energy Index" is created for the ease of computing the energy reduction provided by wall insulation. The creation of the wall simplified energy index is to provide an easy method to estimate energy reduction due to wall U-value selection in Malaysian climate zone and should not be used for any other purpose.

Table 14.2 Energy and peak load impact of wall insulation (medium night time parasitic load scenario)

Descriptions	Ashrae U-value (W/ m^2K)	Wall simplified energy index (kWh/year of m^2 of wall area)	Peak cooling load index (ton/m^2 wall)
Steel sheet, 10 mm	6.68	55	0.476
Concrete wall, 100 mm	3.40	32	0.457
Brick wall, 115 mm	2.82	30	0.456
Double brick wall with 50 mm cavity, 300 mm	1.42	25	0.452
Autoclave lightweight concrete, 100 mm	1.25	24	0.450
Autoclave lightweight concrete, 200 mm	0.75	22	0.449

Table 14.2 is a sample output of insulated wall effects based on a medium night time parasitic load. This table can be used to compute approximate energy reduction and peak load reduction due to the use of wall insulation.

9 Chapter 8 – Roof Insulation

There are many different types of roofing systems in use in Malaysia. The three most commonly practiced roofing systems are: concrete flat roof, light-weight pitched roof and light-weight pitched roof over concrete flat roof. The energy efficiency behaviour of insulation provided on these roof systems is slightly different for each type of roof system and is particularly affected by the different hours of air-conditioning operation in the spaces immediately below the roof.

A typical roof is fully exposed to the entire sky dome and is strongly influenced by the received solar radiation and the effective sky temperature. During daytime, radiation heat transfer on the roof is both being received and reflected at the same time. The sun (solar radiation) is heating up the external roof surface, while the rest of the sky dome (effective sky temperature during daytime averaging around 20 °C) is cooling the external roof surface. The net radiation gain on the roof during daytime will be positive due to significantly higher radiation heat transfer from the sun and the hot external roof surface will transmit heat into the building via conduction through the roof. Insulation of the roof in this case will reduce the heat that is transmitted into the building, reducing the building cooling load.

However, during night time, where there is no solar radiation heat gain and the average effective sky temperature is 15 °C (Chapter 3), heat is removed from the roof surface into the sky via the radiation heat transfer mechanism. Warm temperature from the internal building space below the roof is conducted to the top surface of the roof and rejected to the cold night sky. The provision of insulation in the roof reduces the amount of heat that is rejected out of the roof during night time.

Table 14.3 Energy reduction of flat roof insulation air-conditioning hours: 8 am–5:30 pm

Flat roof insulation (mm)	Energy used kWh/m^2 of roof area per year	RM/m^2 reduction per year
0	135.06	
25	124.19	3.80
50	122.95	4.24
75	122.42	4.42
100	122.12	4.53
200	121.63	4.70
300	121.63	4.70
400	121.42	4.77
500	121.39	4.78

Depending on the hours of operating the air-conditioning system, the provision of insulation in roofing systems can be beneficial or detrimental to the energy efficiency of the building.

Energy simulation was conducted and tables were provided in this chapter to allow selection the insulation appropriate for the building usage. Table 14.3 shows a sample output for flat roof with an air-conditioning hour of 8 am–5:30 pm scenario. The value of Ringgit Malaysia saved per m^2 of roof area is provided to allow a user of the Guideline to make a quick payback analysis based on the current electricity tariff in Malaysia.

10 Chapter 9 – Atrium Ventilation Strategy

It is commonly considered that natural ventilation in atrium spaces is an energy efficient feature in buildings. However, if the atrium space is surrounded by air-conditioned spaces, the exposure of the atrium space to outdoor air will introduce more surface area of the building to be in contact with outdoor air temperature and humidity. This type of exposure increases the risk of conduction of heat and air leakages into the air-conditioned spaces increasing building energy consumption. At the same time, the possibility of air leakages between the conditioned office spaces into the atrium space due to cracks surrounding the windows (or worse, open windows) will help to cool the ground floor of the atrium space because cold and dry air is heavier than hot and humid air and will drop to the bottom of the atrium space. However, if natural ventilation is practiced for the atrium space, these cold air leakages from the offices will be blown away without providing much benefit to the occupants at the atrium space.

Building simulation studies were conducted on ten different options for the atrium ventilation strategies. It was found from the studies that it was not possible to provide comfort conditions at all times if an atrium space is naturally ventilated. At best, comfort condition up to 66 % of the typical working hours is achievable using a combination of natural ventilation and doors/roofs opening hours in the atrium space to provide building energy reduction up to 3.3 %. It was also found that building energy reduction up to 1 % is possible for an air-conditioned atrium scenario when it is naturally ventilated at night. The colder outdoor air during night time cooled the building structure down, resulting in less energy required to cool the atrium space during daytime.

11 Chapter 10 – Zoning and Infiltration Control

11.1 Zoning

There are architects who practice energy efficiency zoning in buildings by locating rooms according to the flow of leakage of air-conditioned air from the coldest room towards the warmest room. This will ensure that leakage of cold air from the coldest room will benefit other spaces before the cold air is allowed to escape out of the building. This assumption is tested in this chapter and the energy saving potential from such an implementation is also provided in this chapter. The result of the study indicates that this is indeed a good design practice. Having 24 h air-conditioned spaces surrounded by 8 h air-conditioned spaces saves up to 2 % energy per floor. The result of the study also indicates that if 24 h air-conditioned spaces are required to be located on the façade, energy lost can be significantly minimised by ensuring that minimum glazing area is used.

11.2 Infiltration

Infiltration of outdoor air into air-conditioned spaces has significant impact on energy efficiency and indoor environmental quality in building. In the Malaysian climate, infiltration introduces both sensible and latent (moisture) heat into building (Razad and Tang 2010). The latent heat (moisture) presents a larger problem for the building because not only does moisture require significant amount of energy to be removed by the air-conditioning system, excessive moisture also leads to a higher risk of mould growth in the building.

Eight simulation cases were developed to study the impact of infiltration, ranging from one open main door, up to 1.6 % windows opened and finally various crack flow coefficients of window perimeters (Technical Note AIVC 44, 1994). It was found with the test model that up to 12 % energy saving was possible by minimising infiltration based on these various parameters. The financial payback study was also provided for improving the crack flow coefficient of window perimeters.

Table 14.4 Energy reduction of moving regularly unoccupied rooms away from daylight area

Percentage working hour individual offices will be empty.	Saving kWh/m^2 per year of moved individual offices.	RM saved per year per m^2 of moved individual offices.
50 %	12.6	4.40
30 %	7.4	2.60
10 %	2.8	0.98

12 Chapter 11 – Interior Layout of Offices

Interior layout of office space influences energy efficiency in a building predominantly due to the effect of maximizing the benefit of daylight harvesting. Individual rooms in offices are normally allocated to higher ranking (senior) staff. These higher ranked staff are typically busy people that attend a lot of meetings and conferences, spending much of the time not in their own office room. In addition, it is also common to allocate the best daylit spaces for higher ranking staff. This type of interior layout will then cause daylight to be provided to unoccupied rooms, while at the same time electric lights are switched on for open space offices that are typically located away from daylight sources.

An energy efficient design alternative is to switch the position of open space offices with the individual offices. This will then ensure that daylight is provided to predominantly occupied open space offices, while individual offices will have their electric lights switched off when not in use.

Table 14.4 shows the results of a simulation study showing the gains to be made from moving individual offices away from the facade (assuming them to have fully daylight when located next to the façade).

Table 14.4 is the key output of Chapter 12, which shows that the energy saved is only 12.6 kWh per year per m^2 of individual office space moved. Assuming that each person is takes up 10 m^2 of space, for each person moved, it would reduce energy use in the building by 126 kWh per year per person if this strategy is implemented.

13 Conclusion

The Passive Design Energy Efficiency Technical Guideline for Malaysia is intended to provide to the Malaysian building industry a simple and unambiguous guideline on energy efficiency in tropical climate building. A host of simulation studies were made by the Component 4 team of the UNDP-JKR, BSEEP project to provide an approximate estimate of energy reduction potential of each passive feature in building. These energy reduction potentials can be used to compute payback studies and internal rate of return based on the investment cost to be made for applying each of these features on a building.

Acknowledgments Acknowledgement is made to Christopher Barry of Pilkington for his contribution to Chapter 6 on Glazing Properties, which is proving to be a very popular chapter, the draft version was downloaded approximately 8000 times in 4 months (end Aug 2012 until early Jan. 2013).

Acknowledgement is also made to Bruce Rowse of Carbonetix, Australia for his contribution to review all the chapters in the Passive Design Guideline, providing very useful tips that helped to make this document clearer and easier to understand by building design professionals.

References

ASHRAE (2009) ASHRAE fundamentals, F15, Fenestration, Table 13

Boyce P, Hunter C, Howlett O (2003) The benefits of daylight through windows. U.S. Dept of Energy, New York State Energy Research & Development Authority, California Energy Commission, Troy

Deringer JJ, Busch JF (1992) Final Report of ASEAN-USAID Building Energy Conservation Project. Energy Analysis Program, Energy and Environment Division, Lawrence Berkeley Laboratory, University of California, Berkeley, pp 7–8

Razad, E.A., Tang, C.K. (2010), Control of moisture & infiltration for advanced energy efficient buildings. In: The tropics, conference on sustainable building South East Asia, 4–6th May 2010, Public Works Department (PWD), Kuala Lumpur, Malaysia

Technical Note AIVC 44, March (1994) An analysis and data summary of the AIVC's numerical database. Air Infiltration and Ventilation Centre

Chapter 15
Residential Thermal Comfort in Tropics – Bunker House

C.K. Tang and Aida Elyana

Abstract The bunker house is an exploration of a residential home design for the tropical climate that avoids the use of an active air-conditioning system on a modern high-standard of living home, based on adaptive thermal comfort model. It is designed to be semi-buried in the earth to maintain a fairly consistent indoor temperature, while maintaining the ability to harvest daylight deep into the home. Instead of attempting to reduce the outdoor air temperature, the bunker house target is to reduce the peak surface temperature on the internal surface of the roof, wall, floor and glazing in an attempt to reduce the operative temperature inside the house. The use of evaporative cooling was not considered for this project because it increases the risk of mould growth due to the higher relative humidity level.

Keywords Bunker house • Humid tropics • Thermal comfort

1 Introduction

The owner of the house was intent to achieve acceptable comfort level with minimum use of an active air-conditioning system. The main concern is to avoid an uncomfortable environment in the house because all the windows are directly facing East. Building energy/thermal simulation studies were conducted to test various passive strategies to reduce the internal surface temperatures for this house. Tests were conducted to find the optimum depth of soil to be used on the roof, the optimum thermal mass properties for the internal walls and external walls, the insulation properties of the internal and external walls, the glazing properties to balance the need for daylight and heat rejection, the external blind operational strategies, natural ventilation strategy and daylight harvesting strategies to negate the use of electric lights during daytime to keep the home cool during a hot day. In addition, strategies to reduce heat gain from equipment in the house were also provided.

C.K. Tang (✉) • A. Elyana
Veritas Environment Sdn Bhd, Kuala Lumpur, Malaysia
e-mail: ck.tang@veritas.com.my

T.H. Karyono et al. (eds.), *Sustainable Building and Built Environments to Mitigate Climate Change in the Tropics*, DOI 10.1007/978-3-319-49601-6_15

The results of the simulation studies indicate that it is possible to reduce the peak surface temperatures of the inner roof (exposed ceiling), the internal and external walls, the floor and the glazing significantly. In addition, the peak air temperature of the internal space was reduced further due to the improvements made. These strategies reduce the operative temperature of the home allowing the occupants to be comfortable without the use of an air-conditioning system. The bunker house will be built in Ipoh, Perak, Malaysia and is expected to start construction in mid-2015.

2 Traditional Architecture for Comfort

The main causes of climatic stress in Malaysia are high temperatures, excessive solar radiation, high humidity and glary environment. Examples of traditional architecture in the tropical climate were analyzed to develop an understanding of traditional methods employed to provide comfort to the building occupants.

2.1 Malay Kampong House

The Malay kampong house reduces solar radiation in the house by the use of large roof overhangs. The exposure of the wall area of the house to direct sunlight is small, reducing the heat gained due to solar radiation (Yuan 1987). The use of timber walls, floors and 'lipah' (leaves) on the roof reduces the thermal mass of the building, allowing the building envelope to cool down quickly during night time (Fig. 15.1).

A large overhanging roof and timber window louvres are provided to reduce glare from the open skies. The use of a ventilated roof space, with openable windows, allows effective ventilation of the building. The house is raised on stilts to allow full cross ventilation below the house. Compound areas are shaded by trees and vegetation to create a cooler microclimate for the house (Yuan 1987).

2.2 Chinese Colonial House

The Chinese colonial shophouses are approximately 20 m deep (Fels 1994). An internal courtyard is placed in the middle to promote ventilation and allow light to penetrate to the rear section of these houses. The verandahway on the street, also known as "five foot" way – "*kaki lima*" shaded the front of the house. As opposed to the traditional Malay kampong houses, the heavy thermal mass of thick brick walls is used to reduce peak temperature in these houses. A series of lightwells brings air and light into the interior spaces of these long shophouses. A roof system with raised mini-gable is provided to release hot air from the roof. Adjustable horizontal louvres

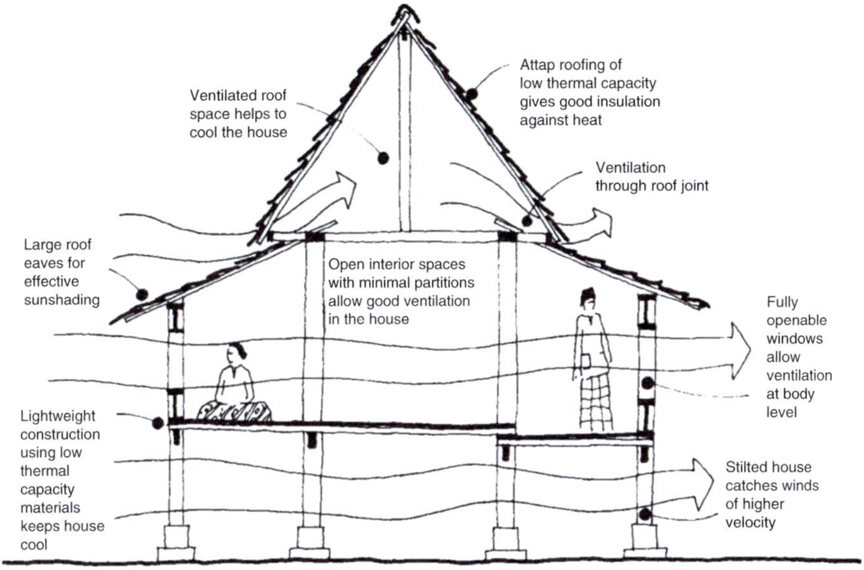

Fig. 15.1 Typical Malay Kampong house

Fig. 15.2 Typical Chinese colonial house

are found in these houses which function to allow natural ventilation, prevent glare from direct sunlight and help to diffuse light into the house (Fig. 15.2).

2.3 *British Colonial Houses*

British colonial houses are typically built with thick walls that provide heavy thermal mass (Ahmad 1997). Louvred windows are provided for daylight harvesting, glare protection and ventilation. Shaped gable roofs are provided for upward flow of air and to function as shading devices to the windows (Fig. 15.3).

Fig. 15.3 Typical British colonial house

2.4 Key Observations

In terms of thermal and visual comfort, all three types of traditional houses in Malaysia shared the following common features:

1. Large roof overhangs to provide shading to the windows to prevent heat gain from the sun.
2. Daylight is harvested into all spaces.
3. Openable windows are provided with the ability for the windows to be fully open or fully closed.
4. Louvred windows are provided to allow control of the amount of daylight harvested.
5. Ventilated roof to reject heat from the roof.

The dissimilarities are found in these areas:

1. Heavy thermal mass is used by the Chinese and British houses, while low thermal mass is used by Malay houses.
2. Malay houses are built on stilts to promote ventilation below the house while the Chinese and British houses are built on the ground.

3 Thermal Comfort and Malaysia Climate

The de Dear Adaptive Thermal Comfort model was used as the basis to improve the comfort for the house. This comfort model indicates that the comfort of a naturally or hybrid ventilated space is directly related to the operative temperature (de Dear et al. 1997). The operative temperature of a space is the combination of two parameters – the air temperature and mean radiant temperature, therefore, in addition to keeping the air temperature low for comfort in a tropical climate, this comfort model says that it is also possible to reduce the peak surface temperature of the roof, wall, floor and glazing in the house to improve thermal comfort.

For a tropical climate such as Malaysia, the average outdoor air temperature is relatively constant throughout the year at 26.9 °C (Tang and Chin 2013) Based on this, the de Dear Adaptive Thermal comfort model says that 90 % of the people will be satisfied with an operative temperature below 28 °C. More importantly, this comfort model indicates that the relative humidity of the space is not a major cause of concern for thermal comfort. The design objective for this house is then to ensure that the air temperature and surface temperatures are kept below 28 °C for the longest duration possible, to ensure that comfort conditions are provided in the house.

4 Thermal Comfort Strategies for Bunker House

The following key thermal comfort strategies were proposed for the project:

1. Reduction of heat gain from the outdoor environment into the house to keep air temperature low.
2. Keep the peak mean radiant temperature of every surface to the minimum to provide lower operative temperature in the house.
3. Reduction of heat produced within the internal spaces of the house.
4. Make use of free cooling provided by the environment.

It is also possible to reduce the indoor air temperature by evaporative cooling methods. However, this option is ruled out because the simulation study made showed that the implementation of such a strategy will cause the relative humidity inside the house to be constantly above 70 %, which would encourage mould growth in the house.

4.1 Façade Optimization

The façade design needs to address the following:

1. reduce the heat gain due to solar radiation from the sun during daytime,
2. allow the harvesting of daylight when the house is occupied,
3. allow solar gain to be excluded when the house is unoccupied and daylight is not needed,
4. allow the house to be open to natural ventilation when weather permits it, and
5. have the ability to open the house to the night sky during night time to benefit from the 15 °C average effective sky temperature at night in Malaysia (Tang and Chin 2013).

In order to meet these criteria, the façade was proposed with full height sliding doors using high performance low-e glazing, to allow visible daylight to be harvested with minimum heat gain. The full height sliding doors will allow the façade to be opened during night time to benefit from natural ventilation and the night sky.

In addition to that, movable external shading with operable louvres was also proposed to provide shading to the glazed facade. The louvres are operable by the building occupants to adjust the amount of glare and daylight harvested into the building. These louvres also allow the building occupants to close them totally to minimize heat gain into the house during daytime when the house is unoccupied, allowing the house to be cooler for the evening hours (when the occupants are back in the house).

During night time, the movable external shading louvres and the sliding doors are moved to the side to expose the inner part of the house to natural ventilation and the night sky. With the average effective night sky temperature in Malaysia of ~15 °C (Tang and Chin 2013), the house would benefit from rejecting heat to the night sky via radiation heat transfer.

4.2 *Daylight Optimization*

In addition to the harvested daylight from the façade of the house, skylights were introduced to bring daylight into the deeper part of the bunker house, for a guest room and a bathroom in the house. The harvested daylight from the roof is designed to be diffused by a horizontal reflector located directly below the skylight. Due to the reason that the house is buried in the soil, there is a possibility of people climbing on to the roof surface. Therefore, in addition to diffusing the daylight, the horizontal reflector is also designed to prevent any direct line of sight of the occupants in the rooms to provide privacy (Fig. 15.4).

Extensive daylight simulation studies was conducted for both the skylight and harvested daylight from the façade to ensure that a minimum daylight factor of 0.5 % is provided for all regularly occupied spaces in the house. In addition, design strategies were put in place to ensure a glare free environment for the occupants of the house. A daylight factor of 0.5 % in the Malaysian climate ensures that a minimum of 100 lux of light is available 80 % of the time from the hours of 8 am to 6 pm (Tang and Chin 2013).

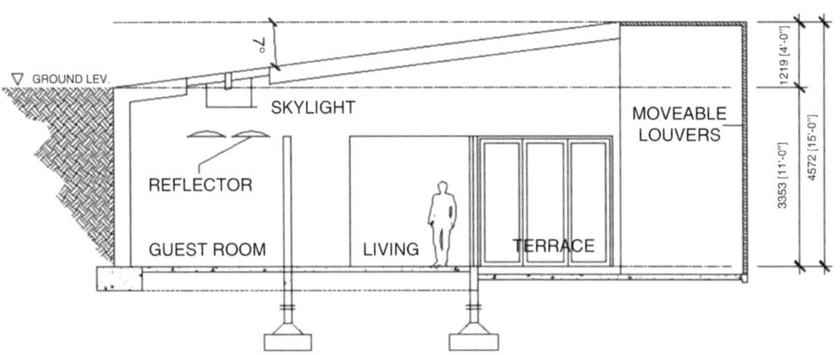

Fig. 15.4 Cross section with proposed daylight strategies

5 Reducing Mean Radiant Temperature

5.1 *Roof Optimization*

The concept for this house is a bunker house that is buried in the soil. Various soil thicknesses on the roof i.e. 100 mm, 200 mm, 300 mm, 400 mm and 500 mm were tested. Figure 15.5 shows that increasing the soil thickness beyond 300 mm provided minor benefits. With 300 mm soil thickness on the roof, the peak internal surface temperature was simulated to be ~9.8 °C lower than the case without any soil on the roof (Figs. 15.6 and 15.7).

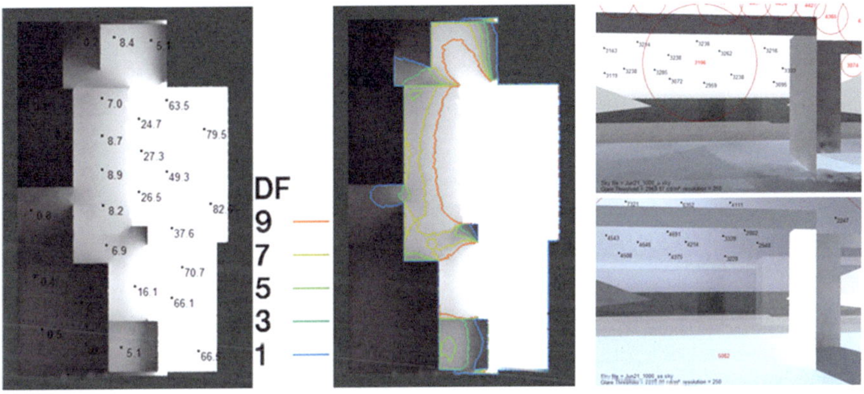

Fig. 15.5 Simulated daylight factor and illuminance intensity without daylight harvesting system

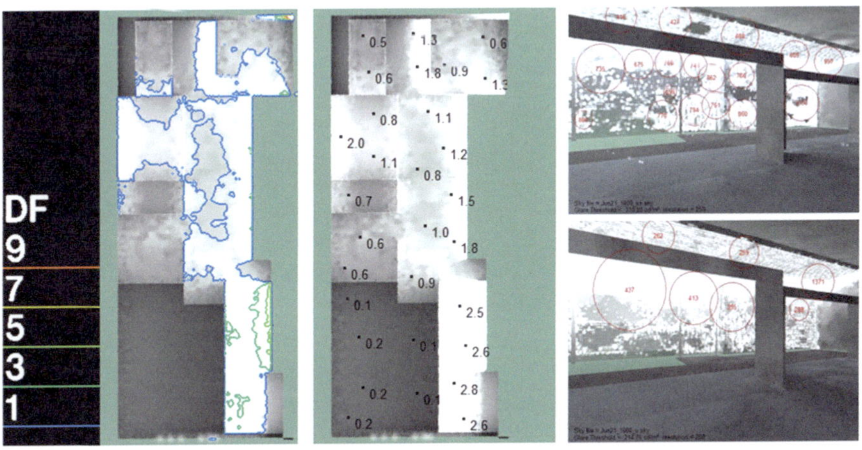

Fig. 15.6 Simulated daylight factor and illuminance Intensity with daylight harvesting system

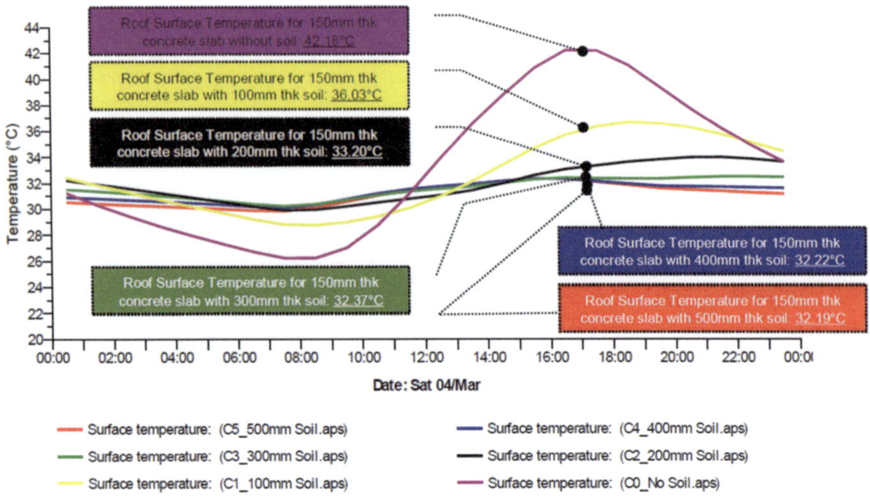

Fig. 15.7 Soil thickness studies on the roof

5.2 Thermal Mass, Increasing Concrete Thickness on the Roof, Floor and Internal Wall

Implementation of 300 mm thick concrete roof slab from the original 150 mm, further reduces the roof peak surface temperature by 1~1.2 °C (Figs. 15.8 and 15.9).

The implementation of 300 mm thick concrete floor slab instead of 150 mm reduces the floor surface temperature up to ~0.5 °Cduring daytime. However, on particularly hot days, it has a small increase of the surface temperature in the late afternoon, due to the received solar radiation (Figs. 15.10 and 15.11).

The implementation of full brick wall of 220 mm thick instead of half brick wall of 110 mm thick, reduces the wall surface temperature by 0.4~0.6 °C during daytime (Figs. 15.12 and 15.13).

5.3 High Performance Double Glazing, Low Emissivity

It was found that the implementation of high performance double glazing with low emissivity brings significant reduction in the surface temperature of the floors and walls, in addition to the reduction of the glazing surface temperature. Figure 15.14 shows the reduction of surface temperature of the floor due to the use of high performance glazing. In addition, Fig. 15.15 shows a reduction of wall surface temperature by ~1 °C due to this implementation. Finally, the glazing surface temperature was shown to reduce by ~6.5 °C in Fig. 15.16.

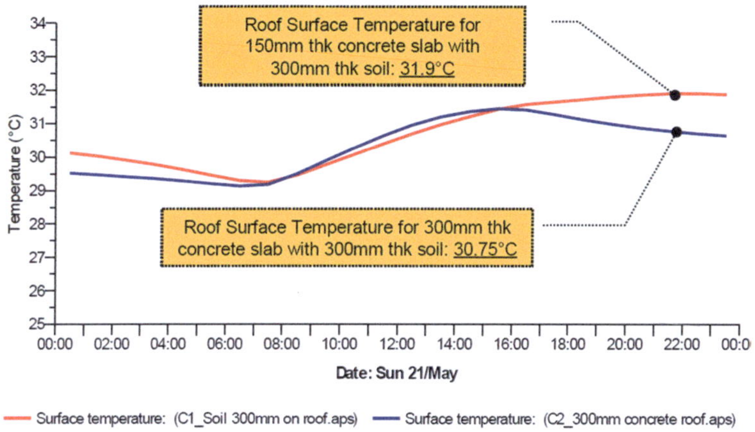

Fig. 15.8 Roof surface temperature (hottest day)

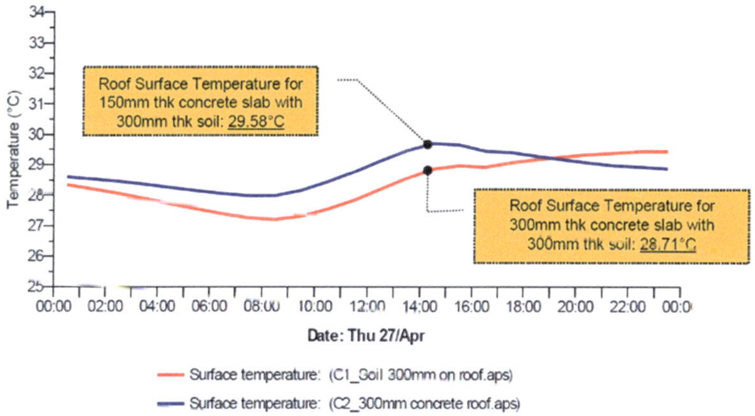

Fig. 15.9 Roof surface temperature (coldest day)

6 Simulation Parameters

Table 15.1 lists the operating parameters used in simulation studies. In addition, all studies were conducted assuming that façade doors and windows are open from the hours of 8 am to 10 am in the morning during the breakfast hours on the terrace space, closed from 10 am to 6 pm to prevent hot air from heating up the indoor spaces and open from 6 pm to 12 midnight to allow the outdoor air to cool down the indoor spaces when the house is occupied.

A test case was conducted where the doors and windows are opened from the hours of 8 am to 6 pm and it was found that in parts of the house, the air temperature increases by ~1.2 °C as shown in Fig. 15.15. This indicates that it is a better strategy to keep the doors and windows closed when the outdoor is air is hot (Fig. 15.17).

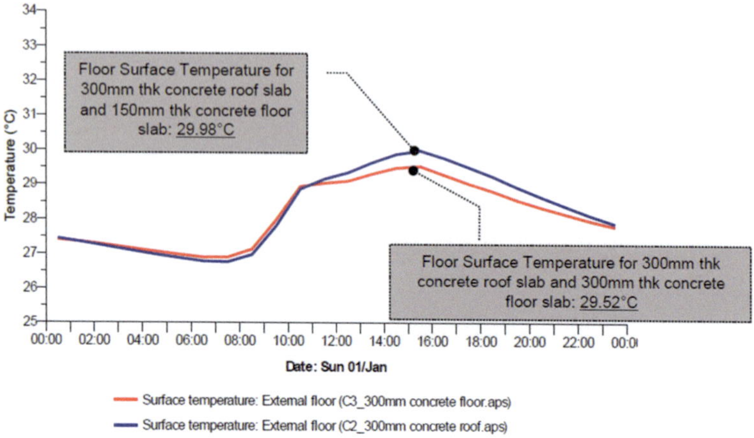

Fig. 15.10 Floor surface temperature (coldest day)

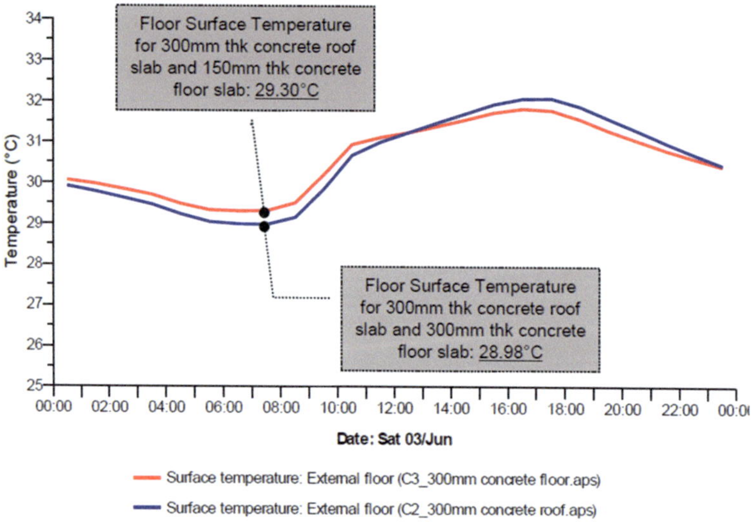

Fig. 15.11 Floor surface temperature (hottest day)

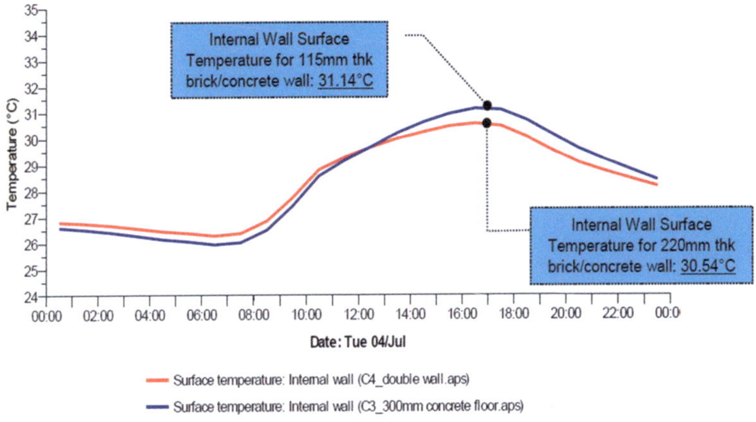

Fig. 15.12 Internal wall surface temperature (coldest day)

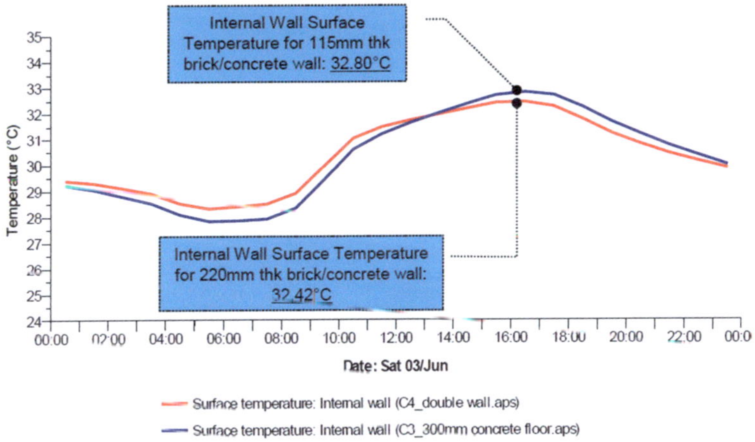

Fig. 15.13 Internal wall surface temperature (hottest day)

7 Summary

In summary, with all proposed implementations carried out, the air temperature of the house was reduced by 2.2 °C, the roof surface temperature is reduced by ~10.4 °C, the floor surface temperature is reduced by ~2.1 °C, the internal wall surface temperature is reduced by 3.2 °C, the external wall surface temperature is reduced by 2.7 °C and the glazing surface temperature reduced by ~7.9 °C. All these temperature reductions combined to provide a significant improvement in thermal comfort as compared to a conventional house in Malaysia.

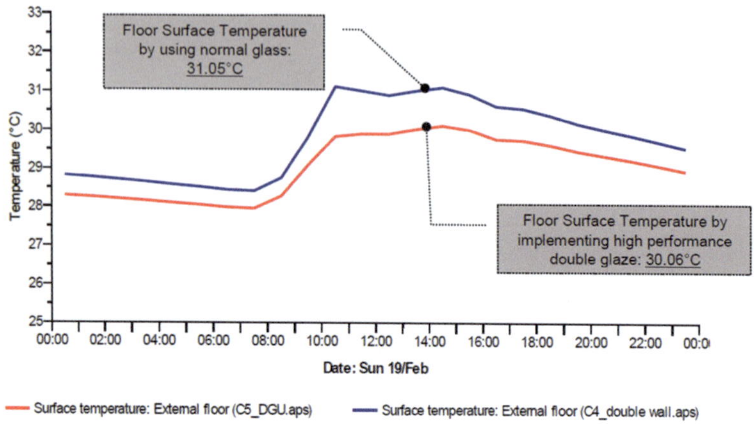

Fig. 15.14 Floor surface temperature

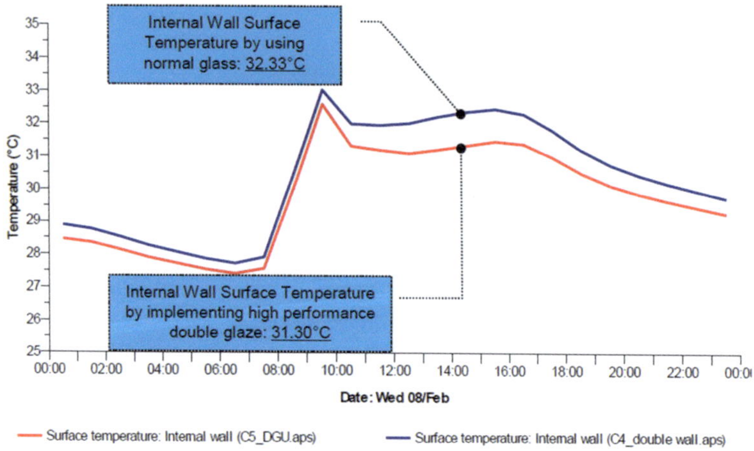

Fig. 15.15 Internal wall surface temperature

Figure 15.18 shows the air temperature simulated for a base case scenario and most comfortable case scenario. During the hottest time of the day, air temperature is reduced by approximately 2.2 °C, down to 31.2 °C, while the average peak mean radiant temperature of the house is reduced by 5.7 °C, down from 35 °C to 29.3 °C. Although these temperatures are above the required operative temperature for 90 % comfort criteria, they only occur during the hottest day of the year based on the simulated test reference year weather data. In short, the hours of discomfort for the entire year are expected to be minimal.

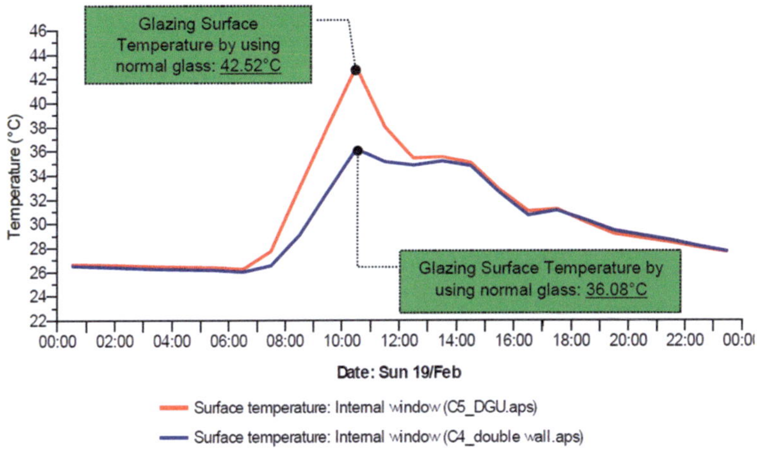

Fig. 15.16 Glazing surface temperature

Table 15.1 Operating strategy implemented in simulation studies

Description	Input data/operation strategy	
Construction material		
Glazing properties	Double glazed low E	
	SC = 0.26	
	U-value 1.9 W/m²K	
External wall properties	300 mm thick concrete wall with 15 mm cement screed	
	U-value 2.8 W/m²K	
Internal wall properties	Internal 220 mm Brick Wall with 15 mm cement screed,	
	U-value 2.5 W/m²K	
Floor properties	300 mm thick concrete slab with 15 mm cement screed	
Roof properties	300 mm concrete flat roof with 300 mm soil on top	
	U-value of 4.5 W/m² K	
Lighting strategy	Lighting power density = 3 W/m²	
Ventilation strategy	Non-air conditioned.	
Small power	Radio	5 W standby 20 W playing
	ASTRO	60 W
	LED TV	100 W
	DVD Player	50 W standby 80 W playing
	Ceiling Fan	85 W (high speed) 25 W (low speed)
Small power	Freezer	350 W
	Refrigerator	400 W
	Expresso (Coffee Maker)	352 W
	Dishwasher	1200 W
	Washer	500 W
	Dryer	4000 W
	Laptop	36 W
	Printer	140 W
	Scanner	16 W
	Exhaust fan	30 W

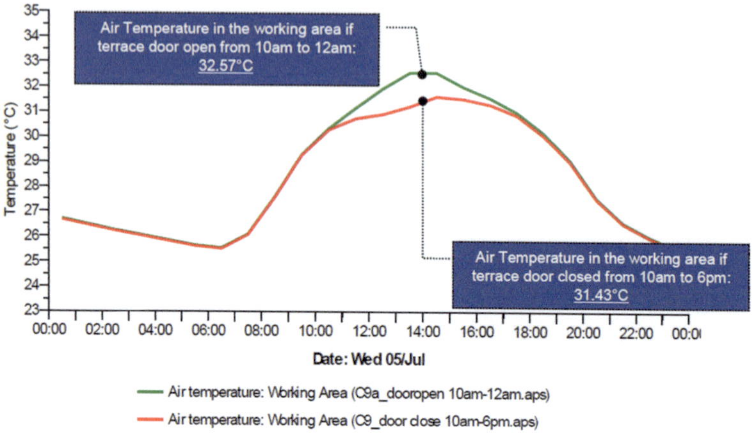

Fig. 15.17 Air temperature in working space

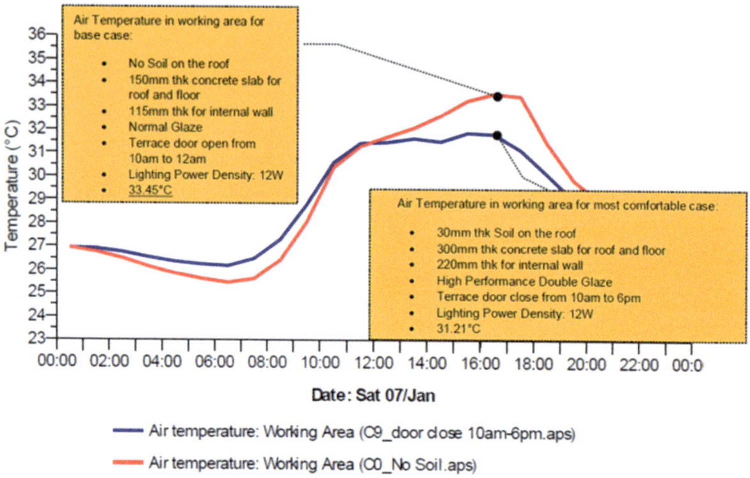

Fig. 15.18 Overall air temperature reduction in working space

References

Ahmad AG (1997) British colonial architecture in Malaysia 1800–1930. Museums Association of Malaysia, Kuala Lumpur

de Dear R, Brager G, Cooper D (1997) ASHRAE RP-884, developing an adaptive model of thermal comfort and preferences. Macquarie Research Ltd, Center for Environmental Design Research, University of California, Berkeley

Fels PT (1994) Penang's shophouses culture. Places 9:46–55

Tang CK, Chin N (2013), Building energy efficiency technical guideline for passive design. Malaysia Building Sector Energy Efficiency Project (BSEEP)

Yuan LJ (1987) The Malay house: rediscovering Malaysia's indigenous shelter system. Malaysia Institut Masyarakat, Pulau Pinang

Chapter 16
Advancing Sustainability in the Tropics – The International School of Kuala Lumpur

C.K. Tang, Julian Saw, and Aida Elyana

Abstract This is a case study about designing a holistically sustainable school in a tropical climate. The school described here is the proposed new International School of Kuala Lumpur (ISKL) located in the heart of the city, Kuala Lumpur, Malaysia. A building is only truly sustainable when all design areas meet sustainable goals. ISKL is designed to be that building, attempting to be sustainable from all design angles. This includes energy efficiency, indoor environmental quality, daylight harvesting, transportation energy, water efficiency, renewable energy, solid waste reduction and permaculture. Design goals and methodology of each strategy are described in this paper.

Keywords Sustainability • School • Energy efficiency • Tropics

1 Introduction

The construction of large buildings is an essential part of any major city's development. With more and more of the world's population migrating from rural to urban areas, it is almost certain that cities will grow in size and density. According to the World Bank, urban population in Malaysia has increased from 67 % in 2007 to 73 % in 2013 and it continues to be on the uptrend (The World Bank 2015). The growth in urban population spurs the demand of more buildings in the city. However, the construction of buildings has an adverse effect on the natural environment. The built environment is a major consumer of raw materials, generates waste, consumes energy and emits greenhouse gases. That is why sustainable construction and green buildings have become an important part of the construction industry in Kuala Lumpur over the past few years.

The proposed new International School of Kuala Lumpur (ISKL) is designed to achieve sustainability in all aspects of design. This article presents how sustainability

C.K. Tang (✉) • J. Saw • A. Elyana
Veritas Environment Sdn Bhd, Kuala Lumpur, Malaysia
e-mail: ck.tang@veritas.com.my

T.H. Karyono et al. (eds.), *Sustainable Building and Built Environments to Mitigate Climate Change in the Tropics*, DOI 10.1007/978-3-319-49601-6_16

can be achieved by providing a methodology and design goals in the areas of energy efficiency, water efficiency, indoor environmental quality, daylight harvesting, transportation energy, renewable energy, solid waste reduction and permaculture. The design of ISKL approaches sustainability in a holistic manner.

ISKL is a proposed international school located at the heart of Kuala Lumpur city in Malaysia. Construction of the building is targeted to be completed in August 2018. The school will have a capacity for 2500 students and includes early years, elementary, middle, and high schools. The whole campus consist of clusters of low rise buildings. This includes a five storey administrative building with library, sports, and academic facilities and a cafeteria. There are also five teaching blocks with three or four storeys each, a gym and a performing arts center.

The following sections outline sustainability goals and methodology for achieving the targets for ISKL.

2 Energy Efficiency

In Malaysia, the Building Energy Intensity (BEI) value is widely used to measure and rate a building's energy consumption. The unit for BEI is kWh/(m^2.year). It is calculated according to the following formula.

$$BEI = [(\text{TBEC} - \text{CPEC} - \text{DCEC}) / (GFA(\text{excluding}\, car\, \text{park}) - DCA - GLA * FVR)] \\ *[52 / WOH]$$

Where:

TBEC: Total Building Energy Consumption (kWh/year)
CPEC: Car park Energy Consumption (kWh/year)
DCEC: Data Centre Energy Consumption (kWh/year)
GFA (excluding car park): Gross Floor Area exclusive of car park area (m^2)
DCA: Data Centre Area (m^2)
GLA: Gross Lettable Area (m^2)
FVR: Weighted Floor Vacancy Rate of GLA (%)
52: Typical weekly operating hours of office buildings in KL/Malaysia (hours/wk)
WOH: Weighted Weekly Operating Hours of GLA exclusive of DCA (hours/wk)

In simple terms, BEI is the total energy consumed per year over the gross floor area of the building. ISKL's BEI target is 35 kWh/(m^2.year). The international average for schools in tropical/sub-tropical climates is 120 kWh/(m^2.year) as published in *Benchmarking Energy Use in Schools* by Sharp (1998). Therefore ISKL aims to consume about 70 % less energy than the international average.

This BEI value has to be achieved by making improvements at every design opportunity available. The simulation software; IES VE, is used to accurately simulate each design improvement and its impact in reducing energy consumption. The list of design improvements are shown below.

2.1 Daylight Implementation

Harnessing natural daylight indoors to reduce consumption of energy from artificial electrical lights. This is presented in greater detail in Sect. 5 below.

2.2 Roof Insulation

Specify 50 mm polystyrene roof insulation with U-value 0.5 W/m^2 K. This reduces heating load on the air conditioning system thus reducing energy consumed.

2.3 Wall Insulation

Specify autoclaved aerated concrete blocks for all external walls for better insulation properties. U-value is 0.9 W/m^2 K. This reduces heating load on the air conditioning system thus reducing energy consumed.

2.4 Glazing Properties

Single glazed low-E glazing is used for all external glazing, with SHGC of 0.34 and U-value of 3.8 W/m^2 K. This reduces heating load on the air conditioning system thus reducing energy consumed.

2.5 Air Tightness

Infiltration introduces both sensible and latent heat into a building in the Malaysian climate. This has significant impact on the energy efficiency and indoor environment quality of the building. Although unwanted infiltration cannot be eliminated entirely, ISKL seeks to reduce it significantly. This is achieved by providing air tightness details and specifications for all external doors and windows. The aim is to reduce infiltration to 0.1 Air Change per Hour (ACH) (Ezzuddin and Tang 2010).

2.6 *Lighting Power Density*

Lighting power is another major source of energy consumption. Light fittings not only consume energy when turned on, lightig also loads the air conditioning system with heat. Lighting power density targets are shown below:

(a) Classrooms and teaching areas: 9 W/m2
(b) Walkways: 7 W/m^2
(c) Administration areas: 9 W/m^2
(d) Toilets: 7 W/m^2
(e) Staircase: 3 W/m^2

2.7 *Air Conditioning Air Side*

A variable air volume (VAV) system is specified. Hence supply air flow can be regulated according to the cooling needs of the zone thus improving efficiency at part loads. Moreover duct design is optimized by reducing bends, tees, transitions, dampers and any others that will add to the AHU fan static pressure. This reduces duct static pressure thus reduces fan power. Total duct static pressure is targeted to be 650 Pa. Fan power is further reduced by using high efficiency filters thus reducing total static pressure to 580 Pa. Overall fan power is reduced by using air foil type fans and IE3 motors. Total fan efficiency is 71.8 %.

A carbon dioxide (CO_2) sensor and control system is also used on the fresh air intake (ASHRAE 2013, Appendix B). CO_2 sensors regulate fresh air intake into the system based on the occupants' need for fresh air. Fresh air although necessary, when drawn into the building is hot and humid, which loads the air conditioning system. The CO_2 sensor is set to 900 parts per million.

A heat recovery system with 50 % efficiency for latent and sensible load is also specified for all teaching blocks. The system recovers energy rejected by the toilet exhaust system and transfers it to the outdoor air intake system located on the roof.

2.8 *Air Conditioning Water Side*

A high chilled water delta-T is specified for both chilled water and condenser water systems. Delta-T is 16 °F for chilled water and 12 °F for condenser water. Moreover pump pressure is also reduced by optimizing pipe size and reducing bends, tees, transitions and any others that will add to the pump pressure. Chilled and condenser water pump pressure is set at 20 m head. The chilled water pumps also use a variable speed motor to vary flow rate according to demand and heat load.

A high efficiency chiller is specified with COP of 6.6. To further improve the chiller performance, a variable speed drive was specified on the chiller. VSD chillers have a variable speed compressor hence have better part load efficiency. Cooling tower efficiency is also improved to 0.0275 kW/HRT.

2.9 Radiant Chilled Slab System

A radiant chilled slab system is used for all classrooms/teaching spaces in the teaching blocks. A radiant cooling system utilizes the floors and ceilings as sources of radiant cooling. Rooms are cooled by circulating water through embedded cross linked polyethylene (PEX) plastic pipes. Such a system is more energy efficient in removing heat from a space because it is using water as the main medium of heat removal as opposed to a conventional air-based system. Water has a thermal capacity of 4.2 kJ/kg.K, it has four times more thermal heat capacity than air which has a thermal capacity of 1.0 kJ/kg.K. Therefore, to provide the same amount of cooling, water requires four times less mass flow rate compared to air.

The chilled slab system is operated in conjunction with the AHU (air handling unit) system. The chilled slab system cannot be used independently because it cannot remove moisture from the air. An AHU system is required to dry the air in the classrooms to prevent risk of condensation on the cold surfaces of these chilled floor slabs. Moreover, the AHU system is required to deliver fresh air to all the classrooms. Approximately 50 % of the cooling will be provided by these AHU units, while the chilled floor slab provides the other 50 % of the cooling requirement.

The cooling of the chilled slabs is done during night time, operating from 1.30 am to 7.30 am. The concrete slabs have high thermal mass, therefore cooling of the system is not immediate. Chilled water that runs through the PEX pipe cools the concrete slabs down to 18 °C. Then the AHU system kicks in at 8 am and runs throughout the day as the building is occupied. The high thermal mass of the concrete slabs allows radiant cooling to take place throughout the day, even though the slabs are not being actively cooled with chilled water during the occupancy period. Simulation results show that the average chilled slab sensible load is 103 W/m^2. Figure 16.1 shows supply air flow rate results simulated for a typical floor. The orange line depicts flow rate for a conventional VAV system, while the red line shows the flow rate for a VAV system coupled with a chilled slab radiant system, which is approximately 50 % of the conventional system.

It is advantageous to charge the chilled slabs at night because commercial electricity tariffs are cheaper. The peak cooling load of the building is reduced because the load is divided into night and day, which will reduce the size of the chiller and all other equipment. Moreover, since both AHU and chilled slab systems operate at similar cooling loads, both can operate using the same chiller, pumps and cooling tower. No additional major equipment is needed.

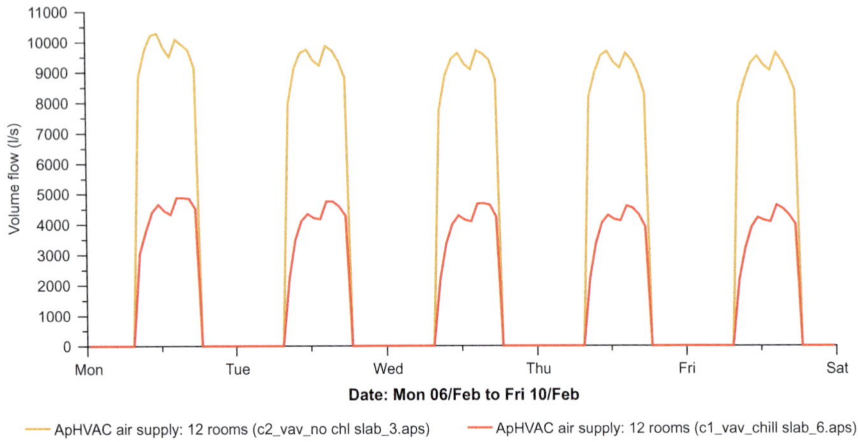

Fig. 16.1 Simulation results showing flow rate of a conventional VAV system in orange compared to a chilled slab/VAV system in *red*

3 Renewable Energy

Renewable energy is provided by photovoltaic solar panels. An expected BEI of 35 kWh/(m^2.year) equates to about 1930 MWh of energy per year. The PV panels will be placed on the roof of the spine block. Available roof space is 5000 m^2, therefore PV panels with a capacity of 250 kWp can be placed on the roof. This array can generate up to 300 MWh of energy per year. Therefore, 15 % of ISKL's energy is currently planned to be provided by renewable energy from photovoltaic panels. If all roof space of ISKL is used, it would be able to provide 100 % of the school energy consumption at ISKL's BEI target of 35 kWh/(m^2.year).

4 Water Efficiency

A number of water efficiency strategies are used to decrease water consumption in ISKL. These strategies are water efficient fittings, rainwater harvesting, grey water recovery, and condensate water recovery. The baseline water demand was estimated for ISKL and shown in Table 16.1. There are four main contributors to water consumption in the school. Combined baseline consumption is estimated to be 92,883 m^3 per year. With the strategies described below, total water demand is reduced by 47 % to 49,489m^3 per year.

4.1 Water Efficient Fittings

The use of high efficiency plumbing fittings is a good method of greatly reducing potable water demand. Water efficient fittings have much lower flow rates compared to conventional fittings while are still able to accomplish their sanitary purposes. Table 16.2 shows annual consumption with conventional fittings (baseline) while Table 16.3 shows annual predicted consumption with efficient fittings. These are

Table 16.1 ISKL's baseline and design water demand

Items	Baseline demand per year (m^3/year)	ISKL's design consumption (m^3/year)
Water fittings	21,838	22,182
Landscape irrigation	46,621	5745
General washing and cleaning	725	725
Cooling tower make-up water	16,253	13,391
Swimming pool	7445	7445
Total	92,883	49,489

Table 16.2 Baseline water fittings demand with conventional fittings

Flush fixture	**Daily use**	**Flow rate (Liters/flush)**	**Duration (Flush)**	**Occupants**	**Water use (Liters)**	**Water use (m^3)**
Water closet (Male)	1	6.00	1	1300	7800	7.80
Water closet (Female)	4	6.00	1	1300	31,200	31.20
Urinal (Male)	3	3.80	1	1300	14,820	14.82
				Subtotal	**53,820**	**53.82**
Flow fixtures	**Daily use**	**Flow rate (Liters/sec)**	**Duration (Sec)**	**Occupants**	**Water use (Liters)**	**Water use (m^3)**
Lavatory faucet	4	0.15	15	2600	23,400	23.40
Café faucet	3	0.15	15	2600	17,550	17.55
Kitchen sink faucet	3	0.15	1800	10	8100	8.10
Showerhead	1	0.21	60	1000	12,600	12.60
Ablution	2	0.15	30	650	5850	5.85
				Subtotal	**67,500**	**67.50**
		Total daily volume			**121,320**	**121.32**
		Annual working days			**180**	**180**
		Total annual volume			**21,837,600**	**21,838**

Table 16.3 Design case water fittings demand with water efficient fittings

Flush fixture	Daily use	Flow rate (Liters/flush)	Duration (Flush)	Occupants	Water use (Liters)	Water use (m^3)
Dual flush WC(Male)	1	4.00	1	1300	5200	5.20
Dual flush WC (Female)	1	4.00	1	1300	5200	5.20
Dual flush WC (Female)	3	3.00	1	1300	11,700	11.70
Urinal (Male)	3	2.00	1	1300	7800	7.80
				Subtotal	**29,900**	**29.90**
Flow fixtures	**Daily use**	**Flow rate (Liters/sec)**	**Duration (Sec)**	**Occupants**	**Water use (Liters)**	**Water use (m^3)**
Lavatory faucet	4	0.032	15	2600	4992	4.99
Café faucet	3	0.032	15	2600	3744	3.74
Kitchen sink faucet	3	0.062	1800	10	3348	3.35
Showerhead	1	0.100	60	1000	6000	6.00
Ablution	2	0.053	30	650	2067	2.07
				Subtotal	**20,151**	**20.15**
		Total daily volume			**50,051**	**50.05**
		Annual working days			**180**	**180**
		Total annual volume			**9,009,180**	**9009**

fittings labelled under the Water Efficiency Labelling Scheme (WELS) from Singapore or equivalent (Water Efficiency Labelling Scheme 2014). As seen in Table 16.3, all fittings have a much lower flow rate compared to conventional fittings. Consumption from water fittings reduces from 21,838 m^3/year to 9009 m^3/year, a reduction of 59 %.

4.2 *Rainwater Harvesting*

Rainwater harvesting is an effective method to reduce potable water usage in a building. Malaysia sees rainfall consistently throughout the year, hence it is good to harness this benefit for ISKL. Harvested rainwater is used for the high landscape irrigation demand. In-house rainwater harvesting simulation software was used to calculate the rainwater tank size. The calculation is done using hourly rainfall data from Meteonorm (Meteotest 2015). Tank size is estimated by measuring hourly rainfall data against daily irrigation needs. For ISKL, rainwater is collected from the roof area only. Rain collected from the roof is diverted via gutters to rainwater downpipes. Rainwater downpipes then divert rain water to the harvesting tank. The

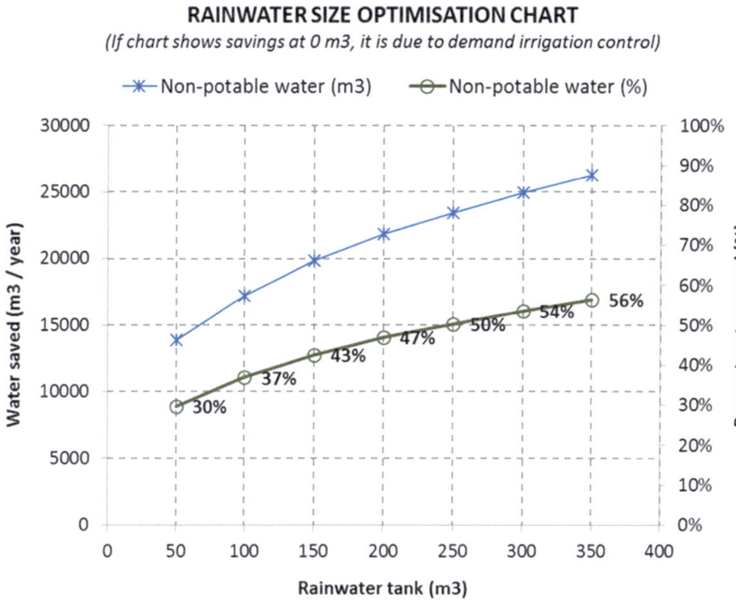

Fig. 16.2 Rainwater harvesting tank chart

rainwater catchment area is 17,893 m^2. A run-off coefficient of 90 % is assumed. This means only 90 % of the rainwater collected is channelled to the tank, the other 10 % is assumed to be splashed out of the building or remains at the catchment area. A first flush system is also used. The first flush system diverts the initial surface rainfall collection. Initial surface rainfall contains a high level of pollutant content. Furthermore, leaves and other debris are often collected from surfaces during the initial stage of rain. Therefore the first 1 mm of rainfall is diverted away from the rainwater tank using the first flush system. This amounts to 16.1 m^3 of initial rainwater being discharged into the storm water drain. All rainfall after that will be channelled into the rainwater harvesting tank.

Figure 16.2 shows the results of the rainwater harvesting software calculations. The chart shows two patterns. Firstly, rainwater collected in a year and secondly, percentage savings of overall irrigation water demand achieved by using collected rainwater. From the chart, we can see that if a tank of 250 m^3 is used, 52 % of the landscape irrigation needs will be met by non-potable rainwater, which is equivalent to a saving of almost 24,439 m^3 of water a year.

4.3 *Grey Water Recycling*

Grey water is collected from all wash basins. This wastewater shall be collected in a tank at the basement and pumped back to serve water closets in the building. Referring to Table 16.3, water collected from all basin faucets amounts to 20.15 m^3 per day. With a recovery ratio of 90 %, total grey water recycled is 18.14 m^3 per day, which is equivalent to 3264 m^3 per year. Therefore a grey water tank of 20 m^3 would reduce yearly wastewater from 9009 m^3 to 5745 m^3, a reduction of 36 %. Recovered grey water will be used to serve all water closets in the building. This further reduces water fittings demand to 5745 m^3 per year.

4.4 Condensate Water Recovery

Condensate water is the water collected from the cooling coils of all AHUs in the building. A typical building would discard this condensate water as waste water. ISKL aims to recover all condensate water as a supplement for the cooling tower make-up water. The calculation below shows how condensate water is estimated.

$$\text{Average Cooling Coil Latent Load, Q} = 680\,kJ\,/\,\text{s}$$

$$\text{Average Cooling Coil Latent Load, Q} = 2{,}448{,}000\,kJ\,/\,\text{h}$$

$$\text{Latent Heat of Vaporization, h} = 2446\,kJ\,/\,kg$$

$$\text{Condensate Water} = \frac{\text{Q}}{\text{h}}$$

$$\text{Condensate Water} = \frac{2{,}448{,}000}{2446}$$

$$\text{Condensate Water} = 993\,kg\,/\,\text{h}$$

$$\text{Condensate Water} = 0.99\,\text{m3}\,/\,\text{h}$$

$$\text{Condensate Water} = 15.9\ \text{m3}\,/\,day$$

Therefore it is estimated that 15.9 m^3 of condensate water is collected per day. The average condensate water collected in a year is 2862 m^3 which reduces cooling tower water demand by 17 % to 13,391 m^3 per year.

5 Daylight Harvesting

Artificial lighting in a building both consumes electricity and is also a source of heat gain on the air conditioning system. Therefore it is worthwhile to harness natural daylight wherever possible. A daylight harvesting strategy was implemented as part of the design. This includes daylight sensors at the building perimeter of each floor, and a window design to draw daylight deeper into the office area while preventing glare. The window design consists of horizontal venetian blinds and light shelves as shown in Fig. 16.3.

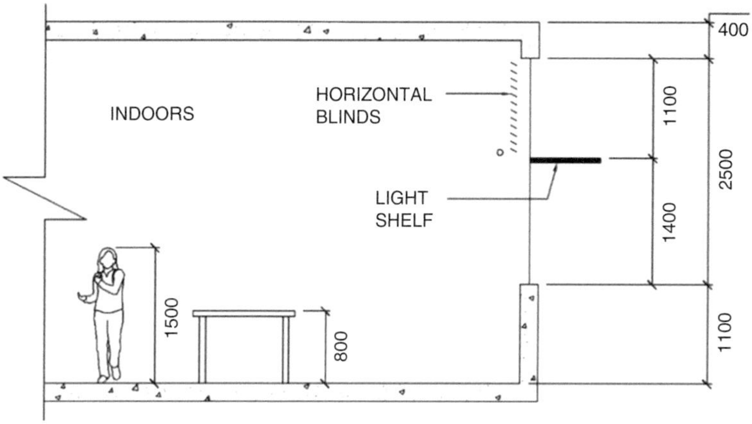

Fig. 16.3 Cross section of the window design for classrooms

Quality of natural indoor daylight is measured by Daylight Factor (DF). Daylight Factor is a measure of a reference point of the available daylight indoors versus the available daylight outdoors during an overcast sky condition. DF criteria used for ISKL are in the range of 0.5–3.5 % for classrooms (Tang and Chin 2013). The daylight simulation software Radiance is used to simulate indoor daylight conditions. Due to the space limitation, it is not possible to present here all results from the simulation but results from one teaching block are shown in Fig. 16.4. This chart shows that for Level 4, natural daylight can light up to 7 m of the classroom from the perimeter of the building. This is nearly the whole width of the room. Therefore artificial electric lights is not needed. For Level 3, up to 6 m is available, for Level 2, 4.5 m and for Level 1, 3 m. Hence about 65 % of the classroom's artificial electrical lights can be switched off.

6 Indoor Environmental Quality

ISKL is designed to meet the requirements of ASHRAE 55 and ASHRAE 62.1. These standards provide guidelines for thermal and environmental conditions for human occupancy and also acceptable indoor air quality. Moreover, with the use of daylight simulation, ISKL is able to provide a more conducive indoor environment with natural daylight without glare.

7 Transportation Energy

A holistic approach to sustainability includes consideration of transportation energy consumed by occupants of ISKL. Hence a methodology to calculate the transportation carbon emissions should be clearly defined and documented. This methodology

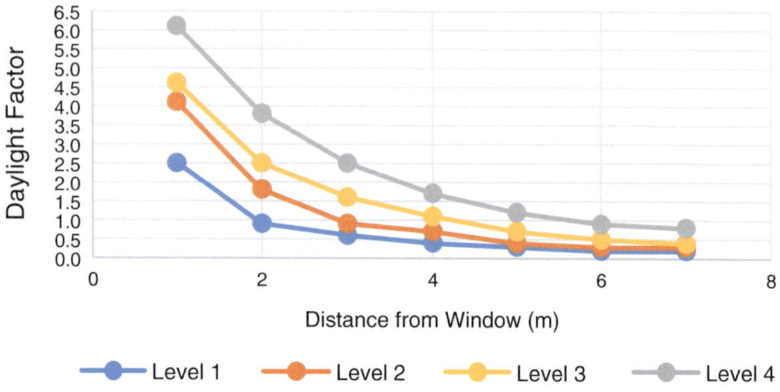

Fig. 16.4 Daylight Factor results for a teaching block

shall also provide a transportation carbon emission performance indicator to specify the carbon emission per student based on the type of transportation used. The performance indicator provides a measurable way to show how well the school is doing in managing transportation energy and provide a way forward to for corrections and improvements.

Below is an outline of the methodology to measure and specify transportation energy in ISKL. The method should be simple and practical to be applied during the operation of the school. The performance indicator shall be given by TCEI (School Transportation Carbon Emission Index per year per student served). This formula provides the transportation carbon emission per student served per year for a school. Ideally, this number should be as close to zero as possible for a school to be truly sustainable. This can be achieved by reducing the carbon emission of individual private vehicles, buses and public transportation and/or by increasing the number of students served by each vehicle travelling to a school.

$$TCEI = \frac{\sum P.Vehicle + \sum S.Bus + \sum P.Transportation}{StY}$$

Where,

TCEI = School Transportation Carbon Emission Index per year per student served.
P.Vehicle = Carbon Emission per year of Private Vehicles.
S.Bus = Carbon Emission per year of School Buses.
P.Transportation = Carbon Emission per year of Public Transportation.
StY=Average number of students per year

The design team will work with the school to implement a system to collect necessary data during operation of the school. Data of carbon emitted from private vehicles will be collected from those that travel to the school using private vehicles. The data shall include type of vehicle, distance travelled, trips per week and the rated CO_2 emission of the vehicle. Similarly, this data will also be collected for school buses. However, there is not much data provided by the Malaysian transport authority regarding CO_2 emission for public transport vehicles. Therefore the design team shall be working on gathering this data before the school is completed.

8 Solid Waste Reduction

ISKL's goal is to reduce waste to landfills by 90 % (The International School of Kuala Lumpur 2015). Waste such as paper, plastic, aluminium, glass and used cooking oil can be easily segregated and recycled, as there are facilities in Kuala Lumpur to collect and recycle these items. Recycling bins shall be provided at strategic locations on the campus to encourage occupants to segregate waste accordingly. However, organic waste from the kitchen and landscape would still be required to be discarded to the landfills as Kuala Lumpur lacks the facilities to recycle such waste today. Therefore ISKL will include an anaerobic digester system to treat organic waste. These strategies will be applied to achieve the goal of 90 % waste reduction. Moreover, the anaerobic digester serves as an educational demonstration tool to students in the school. The anaerobic digester is basically a system with microorganisms that break down organic waste in the absence of oxygen. The process then produces a biogas containing methane, carbon dioxide and traces of other gases that can be used as fuel. Another by-product of the process is a nutrient rich digestate that can be used as fertilizer.

Figure 16.5 shows a concept schematic of the anaerobic digester. The components are as follows:

Kitchen Sink and Food Waste Dispenser Organic food waste from the kitchen is disposed into the food waste dispenser (grinder) and flushed down with rain water or grey water, as a slurry mix into a holding tank below the sink.

Holding Tank, Pump and Check Valve The holding tank, pump and check valve can be omitted if the kitchen sink is located higher than the anaerobic digester tank. If it

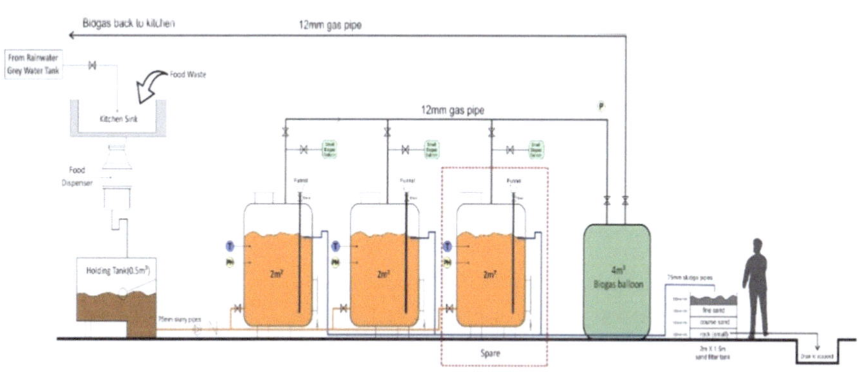

Fig. 16.5 Concept schematic of the anaerobic digester in ISKL

is not possible, the holding tank will collect the dispensed food waste for a temporary period before pumping the slurry into the anaerobic digester tank.

Anaerobic Digester Tank The tank is a standard polyethylene septic tank (UV resistant) that is air-tight, low-maintenance and low-cost. The slurry from the holding tank is piped into the bottom of the anaerobic tank. The tank is sized to allow 20–30 days of digestion period. The maximum gas pressure in the tank is estimated to be less than 2000 Pa (200 mm of water). The top of the tank is fitted with a gas pipe to route the accumulated gas to a biogas balloon. Digested sludge will be discharged from the top of the tank and piped to a sand filter system.

Biogas Balloon The biogas storage balloon is fabricated out of high tenacity polyamide fabric matrix, impregnated with compatible polymer on the inside for the biogas stored and on the outside with Hypolon (or equivalent) for weather and UV resistance. These balloons are collapsible, light weight and can be suspended from the ceiling or walls or placed on the floor.

Below are the advantages for having an anaerobic digester:

(a) Recycles all organic waste generated by ISKL.
(b) Generates methane gas to be used in the kitchen. However the quantum of generation is quite small compared to overall kitchen needs. It is estimated that the system provides 1–3 % of total kitchen needs.
(c) After a digestion period of 30 days, the system will produce about 150–200 kg of digestate a day at steady state operation. This digestate is high quality, nutrition rich top soil that can be used within the landscape area of the school, or bagged up and sold to nearby orchards.

9 Urban Farm and Permaculture

There are many benefits to urban farming (Mobbs 2002). In a school like ISKL, it can be used as a tool to educate students about local food sources, the science of plants, and the process of farming. Moreover, urban farms are a great solution to reduce the cost of transportation of foodstuffs, hence reducing the carbon footprint associated with food production. In the case of ISKL, urban farms within the campus may not produce enough food to supply the needs of the whole campus. But the goal is to introduce the concept of urban farming to the students and community.

Permaculture on the other hand is a concept of "ecological" design that blends various eco-systems together in a sustainable manner, forming an urban farm modelled after a natural ecosystem. The following are plans to be implemented in ISKL.

(a) Vegetable garden on the rooftop with soil depth of approximately 0.5 m depth to maintain a healthy vegetable garden.
(b) Introducing sting-less bees. Bees are one of nature's most important pollinators. Many food plants such as soursop, celery, papaya, cabbage, cauliflower, cucumber etc. require bees to help with pollination. However, to ensure that the bees

remain at the campus, flowering plants with high nectar content are required to be planted to keep them at the site. The bee hives in wooden or clay boxes should be kept at a high position out of reach from school children. Harvesting of honey from the hives can be done via a ladder by the farmer at regular intervals.

(c) Fruit trees such as soursop, rambutan, papaya, or chiku trees will serve as ideal perimeter trees around the campus. When the fruit trees start to fruit, it will create an interesting learning opportunity about tropical fruit trees for the students in the school

(d) Flowering plants shall be planted between the fruit trees as a food source for the bees to ensure that they stay within the campus to help with the pollination.

(e) A variety of crops will be planted each season, with crop rotation constantly conducted to maintain the quality of the soil.

10 Conclusion

The new ISKL campus strives to be a holistically sustainable school. A summary of the design goals and methodology of this holistic approach to meet the important sustainability targets is presented below:

1. By making small improvements at every design opportunity, the building can be made very efficient. The BEI design target is 35 kWh/(m^2.year), about 70 % less energy than the international average.
2. Photovoltaic solar panels with a capacity of 250 kWp will provide 15 % of the annual energy consumption.
3. With the use of water efficient fittings, and implementing rainwater harvesting, grey water and condensate recovery systems, water consumption is reduced by 47 % from a baseline of 92,883 m^3/year to 49,488m^3/year.
4. Daylight harvesting system provides excellent indoor daylight. Up to 65 % of artificial lights can be switched off.
5. A transportation energy plan is developed to provide a measurable way of showing how well the school is doing in managing transportation energy and provide a way forward for corrections and improvements.
6. With a recycling program and an anaerobic digester, a target of 90 % of solid waste shall be recycled.
7. Urban farming and permaculture shall be designed as part of the school as an education tool and a proof of concept for agriculture in the city.

As the project continues from design to construction and then to operation, it is important the design intent continues as well. Strategies will be put in place to ensure all sustainability targets are met throughout the development of the school.

References

ASHRAE (2013) Appendix B, summary of selected air quality guidelines. In: ASHRAE standard 62.1–2013. ASHRAE, Atlandta

Ezzuddin RA, Tang CK (2010) Control of moisture & infiltration for advanced energy efficiency buildings in the tropics. Conference on sustainable buildings South East Asia. Public Works Department, Kuala Lumpur

Meteotest (2015) Meteonorm. Retrieved from Meteonorm: http://meteonorm.com/en/buy/meteonorm_dataset

Mobbs P (2002) Grow your own food. Free Range Practice Guide, p 4

Sharp TR (1998) Benchmarking energy use in schools. ACEEE summer study on energy efficiency in buildings. American Council for an Energy-Efficient Economy, Washington DC

Tang CK, Chin N (2013) Daylight harvesting. In: Tang C, Chin N (eds) Passive design. Building Sector Energy Efficiency Project (BSEEP), Kuala Lumpur, p 76

The International School of Kuala Lumpur (2015) Waste recovery & recycling. Retrieved from Sustainable ISKL: https://sites.google.com/a/iskl.edu.my/sustainable-iskl/

The World Bank (2015) Urban population. Retrieved from World Bank: http://data.worldbank.org/indicator/SP.URB.TOTL.IN.ZS

Water Efficiency Labelling Scheme (2014) Retrieved from List of Products: https://app.pub.gov.sg/WELSSP/ListOfProducts.aspx

Part IV
Plants and Low Carbon Emissions

Chapter 17
Plant Selection and Placement Criteria for Landscape Design

Chun Liang Tan, Nyuk Hien Wong, and Steve Kardinal Jusuf

Abstract This study explores how landscape design can be optimized by considering specific plant traits and their corresponding temperature reduction potential. An initial study was conducted with the aim of quantifying the impact of rooftop greenery on mean radiant temperature (T_{mrt}). Results show that under clear sky conditions, plots with vegetation can reduce surrounding T_{mrt} by up to 6.0 °C. The effect in temperature reduction is evident for a distance up to 3.0 m away from the center of the green plots. Thereafter, a second set of measurements was made to identify specific plant traits that contribute to temperature reduction. Results indicate that the temperature reduction potential of different types of vegetation varies according to their physical characteristics as well as physiological attributes such as plant evapotranspiration rate and shrub albedo. An empirical model was developed to establish the relationship between T_{mrt} reduction, plant evapotranspiration and shrub albedo. Findings from these studies are used as a basis to formulate a framework for landscape planning and design. In the proposed framework, vegetation as well as building information are superimposed using a Geographical Information Systems (GIS) platform. A hypothetical scenario is used to illustrate the efficacy of the proposed landscape planning framework.

Keywords Mean radiant temperature • Outdoor thermal comfort • Rooftop greenery • Vertical greenery

---

C.L. Tan • N.H. Wong
Department of Building, School of Design and Environment, National University of Singapore, Singapore, Singapore

S.K. Jusuf (✉)
Sustainable Infrastructural Engineering (Building Services), Singapore Institute of Technology, Singapore, Singapore
e-mail: stevekj@singaporetech.edu.sg

T.H. Karyono et al. (eds.), *Sustainable Building and Built Environments to Mitigate Climate Change in the Tropics*, DOI 10.1007/978-3-319-49601-6_17

1 Introduction

The cooling effect of greenery has been well documented worldwide (Gao 1993; Ca et al. 1998; Shashua-Bar and Hoffman 2000; Jonsson 2004; Wong and Chen 2005). Recognising the positive impacts of urban vegetation, many governments have initiated policies aiming to improve the condition of the city through the addition of greenery (Beatley 2000; Ra 2006; Zhao 2011). Provision of Urban Green Spaces (UGS) acts as urban lungs, helping to absorb pollutants and releasing oxygen, providing clean air, water and soil and cleansing the city's environment (Hough 1984).

Many studies have been conducted to quantify the benefits of urban greenery. As ground level space is a limiting factor for most urbanized areas, rooftop greenery and sky terraces have gradually become more prominent in the urban landscape. Research into green roofs often focuses on the reduction of roof surface temperature due to the presence of greenery (Wong et al. 2003; Cheng et al. 2010; Perini et al. 2011). Vegetation can reduce the impact of the Urban Heat Island (UHI) effect by shading heat-absorbing surfaces and cooling the air through evapotranspiration (McPherson 1994). The reduction in temperature can lead to lower cooling loads for the building interior (Wong et al. 2009; Pérez et al. 2011). There are also studies into various aspects of rooftop greenery such as the types of plants used, growth substrates, acoustic performance, air quality and maintainability (Akbari 2002; Parizotto and Lamberts 2011; Baik et al. 2012; Saadatian et al. 2013). Various feasibility studies have also been conducted to determine structural and logistical considerations for green roof implementation (Castleton et al. 2010).

The cooling effect exhibited by plants is a result of their metabolic processes, such as photosynthesis and evapotranspiration. The extent to which plants engage in these processes is directly related to the amount of green matter, usually found in the leaves of the plant (Jones 1992). However, this knowledge is not being utilized to its full potential in the current landscape design and planning processes. More often than not, designers assume the cooling effect of plants to be homogenous, resulting in indiscriminate selection and allocation of plants to the landscaped area.

This study seeks to assess the effects of rooftop greenery on mean radiant temperature, as well as to explore the possibility of formulating objective landscape design principles based on the cooling potential of specific plant traits in the tropical urban environment.

2 Methodology

2.1 Rooftop Greenery Measurement

In the initial study, a total of 28 points were set up for mean radiant temperature (T_{mrt}) measurement. Each measurement point, which consisted of a customized globe thermometer fixed at 1.3 m above ground (Fig. 17.1.) and an air temperature

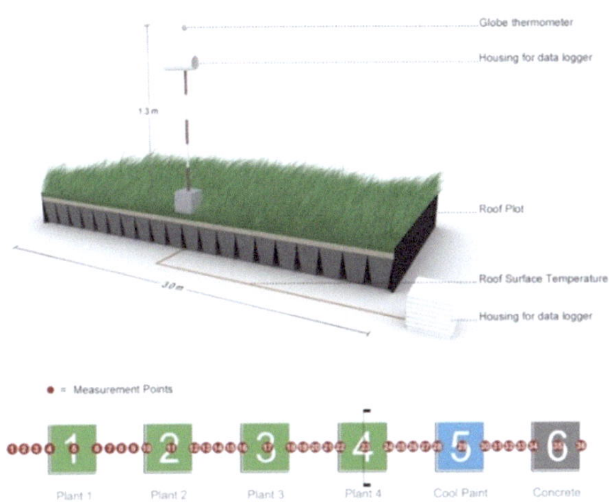

Fig. 17.1 Measurement points and sectional perspective of roof garden setup

Table 17.1 Plot characteristics for rooftop garden

Plot	Characteristic	Specification
1	Shrub	Phyllanthus cochinchinensis
2	Shrub	Heliconia 'American Dwarf'
3	Shrub	*Sphagneticola trilobata*
4	Turf	*Axonopus compressus*
5	Cool Paint	*JOTUN* Jotashield extreme
6	Concrete	Control

sensor housed in a PVC pipe, was secured with a concrete footing. Measurement points were aligned and named as shown in Fig. 17.1. The experiment was conducted at the National University of Singapore SDE1 rooftop. The T_{mrt} of six plots were measured. Each plot had a dimension of 3.0 m by 3.0 m. The first four plots were plots with vegetation. The fifth plot was covered with an acrylic sheet painted with cool paint. The sixth plot was bare concrete, used as a control for the measurement. Characteristics of each plot are shown in Table 17.1. Leaf reflectivity was measured for all plant species used in this study. Leaf total reflectance was measured using a Spectrophotometer (Shimadzu UV-3150 UV-VIS-NIR) for the range of 190–3200 nm.

Each plot was placed at regular intervals of 3.0 m to minimize interference from neighbouring plots. Roof surface temperature was measured underneath Plots 1–4, as well as on Plots 5 and 6 using a Yokogawa multi-logger. Sensors were deployed to measure T_{mrt} of all plots. Measurement was made at 1 min intervals. A total of 42 sensors were deployed. Each globe temperature sensor was attached to a survey pole and measured T_{globe} at 1.3 m above the plots. Six sensors were placed underneath each plot to measure the roof surface temperature. Estimation of T_{mrt} can be done using the globe thermometer (Vernon 1932). Initially developed for indoor usage, the globe thermometer has since been adapted for outdoor use (Nikolopoulou and Lykoudis 2006). The Vernon globe is a 150 mm diameter copper sphere painted black with a thermometer positioned in the middle of the sphere. For outdoor measurement, the 38 mm globe thermometer is a common option as the globe used is a table tennis ball, which can be readily purchased and conveniently replaced (Humphreys 1977). Accuracy of the 38 mm globe thermometer can be adjusted to cater for outdoor conditions by recalibrating the mean convection coefficient (Thorsson et al. 2007). In this study, T_{mrt} was estimated using the following formula specifically calibrated for tropical outdoor use (Tan et al. 2013):

$$Tmrt = \left[\left(Tg + 273.15\right)^4 + \frac{2.20 \times 10^8\, Va^{0.119}}{\varepsilon D^{0.4}} \times \left(Tg - Ta\right) \right]^{0.25} - 273.15 \quad (17.1)$$

Where,

T_g = Globe temperature (°C)
V_a = Air velocity (ms^{-1})
T_a= Air temperature (°C)
D = Globe diameter (m)
ε = Globe emissivity

3 Results and Discussion

Diurnal T_{mrt} profile for measurement points at the center of the plots are shown in Fig. 17.2. The maximum and minimum values of T_{mrt} recorded were 63.0 °C and 24.9 °C respectively. The maximum difference during the hottest time (14:00 h) was

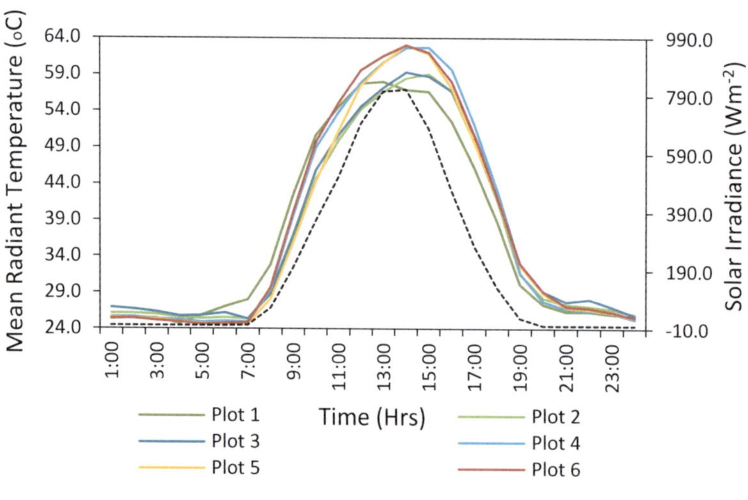

Fig. 17.2 Diurnal T_{mrt} profile for clear sky conditions

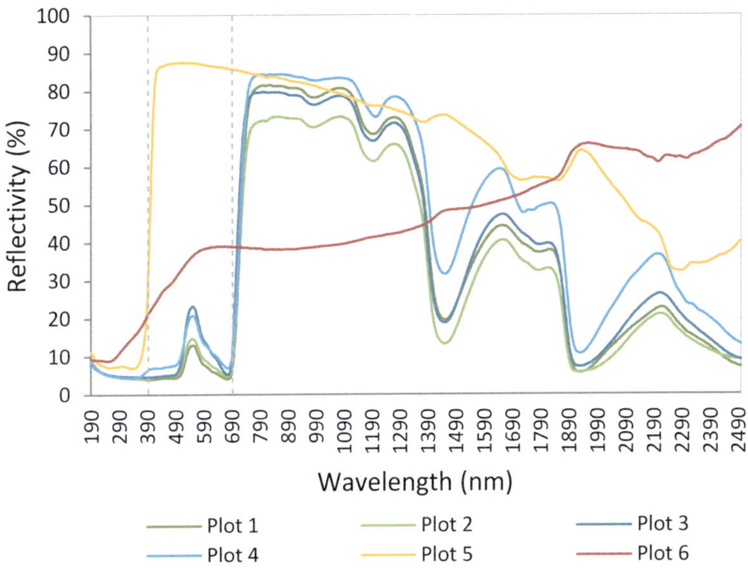

Fig. 17.3 Spectrometer test results

approximately 6.0 °C. It is observed that Plots 4, 5 and 6 exhibit similar T_{mrt} profiles, peaking at approximately 63.0 °C. Of the six plots, Plot 1 (Phyllanthus cochinchinensis), had the lowest diurnal T_{mrt} profile, followed by Plot 2 (Heliconia American Dwarf) and Plot 3 (Sphagneticola trilobata). In the absence of sunlight (01:00–07:00 h and 19:00–00:00 h), the T_{mrt} profile for all six plots remained stable without much fluctuation.

Material reflectivity was measured using a spectrophotometer and results are shown in Fig. 17.3.

Under the visible light spectrum, which spans from approximately 380 to 700 nm, cool paint exhibits the highest reflectivity of more than 80.0%. This is followed by concrete with a peak reflectivity of 40.4%. Vegetation under visible light displays relatively lower reflectivity, peaking at 25.2% (Plot 4), 21.6% (Plot 3), 14.5% (Plot 1) and 13.8% (Plot 2) at around 550 nm. For wavelengths in the near-infrared range (700–2500 nm), the reflectivity of cool paint reduces gradually, while the reflectivity of concrete increases at a similar pace. Reflectivity of the plants increases significantly from 700 to 1400 nm and undergoes a series of fluctuations, rising again from 1600 nm to 1800 nm and 2200 nm. It can be seen that the reflectivity of plants from the range of 701 to 1300 nm can equal that of cool paint.

The diurnal roof surface temperature (T_s) profile is shown in Fig. 17.4. The solar irradiance peaked at 1004.4 Wm^{-2} at 14:00 h. The surface temperature under Plots 1–4 was significantly lower than Plots 5 and 6 during daytime. The maximum difference between Plot 6 and Plot 3 was approximately 14.4 °C. The surface temperature of Plots 1–4 was maintained at between 26.0 and 29.0 °C throughout the day. In contrast, the surface temperature of Plots 5 and 6 increased greatly during the

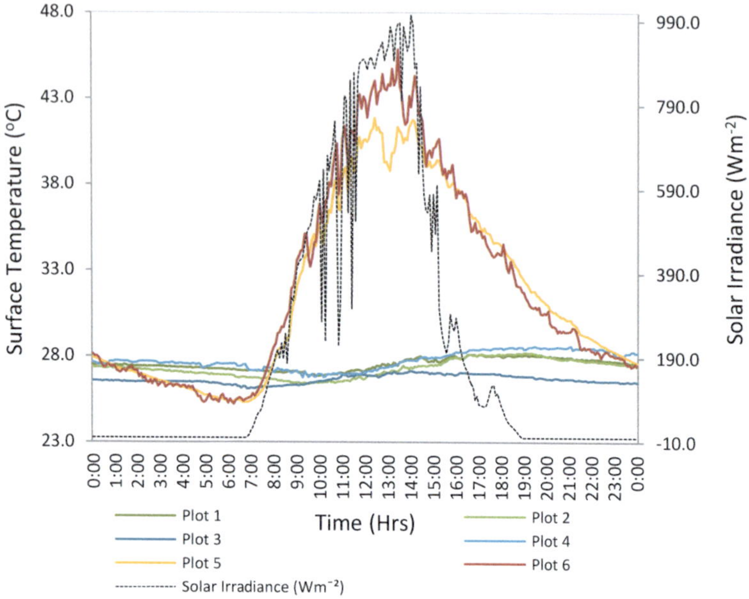

Fig. 17.4 Roof surface temperature – 9th October 2012

day. A peak of 45.9 °C was observed at 13:25 h for Plot 6, while the corresponding temperature of Plots 1–4 is only in the range of 27.0–27.5 °C. It can be observed that the diurnal surface temperature profiles for Plots 5 and 6 were similar for large parts of the day, except for periods of high solar irradiance (11:00–14:00 h), where temperature readings for Plot 5 were slightly lower than those of Plot 6. From 03:00 to 07:00 h, it can be observed that Plots 5 and 6 (Cool paint and concrete roof) had a slightly lower temperature compared to the other plots.

Infrared thermal images were taken on separate days with clear sky conditions. A total of three readings are obtained and averaged. Figure 17.5 shows that when the roof is exposed to direct solar irradiance, the surface temperature of the bare roof can reach up to 63.5 °C. With the exception of Plot 4, all plots with vegetation displayed temperatures lower than the cool roof and concrete roof.

Results show that when cool roofs and green roofs are exposed to direct solar irradiance, the surface temperature is significantly lower compared to the exposed concrete roof. However, only green roofs provide a substantial reduction in mean radiant temperature. At peak solar irradiance, T_{mrt} at 1.3 m above the green roof plots can be up to 6.0 °C lower than above the concrete surface. Besides lowering T_{mrt}, the introduction of green roofs can also help to minimize temperature fluctuations. Fluctuations in radiant temperature will drastically change the radiation absorbed by an individual and the energy budget of the individual, affecting the overall thermal comfort.

Results from the reflectivity test show that for the visible light range (380–700 nm), the cool roof can have a reflectivity of more than 80%. In comparison, the

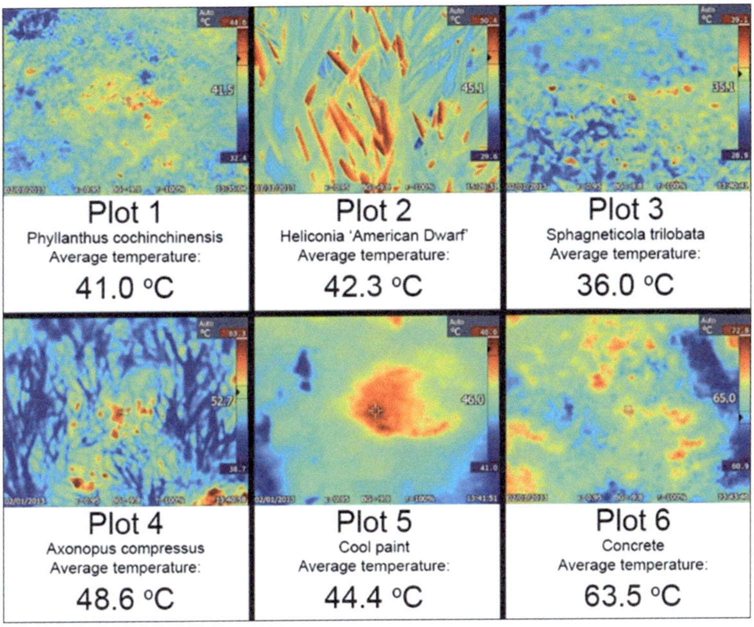

Fig. 17.5 Average roof surface temperature.

reflectivity of plants is in the range of 20%, which is rather low. However, at the near-infrared range, especially from the range of 700 to 1400 nm, the reflectivity of plants can almost equal that of cool paint (Fig. 17.3).

Since up to 50% of the solar irradiance distribution is in the near-infrared zone, this attribute may be crucial for reflecting heat back to the atmosphere. The high reflectivity of plants, in addition to the inherent cooling potential through evapotranspiration, may be the main factors leading to significantly lower T_s, T_a and T_{mrt} values.

3.1 *Landscape Planning Framework*

Numerous studies have highlighted the significance of strategic plant placement in landscape design (Gómez-Muñoz et al. 2010; Simpson and McPherson 1996). Cameron et al. (2014) showed that different plant species varied distinctively in their cooling capacity and the mechanisms for cooling varied between species.

With the benefits of greenery widely acknowledged, the next challenge is to translate this knowledge into industry practice. While there are existing frameworks to objectify landscape planning processes such as maintainability and irrigation (CTLA 1992), methodologies for assessing landscape proposals in terms of plant cooling potential have yet to be formalized in a comprehensive manner. The problem of visualizing the thermal impact of any landscape design is exacerbated by the inclusion of alternate forms of vegetation such as vertical and rooftop greenery.

Recent developments in modeling the outdoor thermal environment have enabled scientists to understand the impact of vegetation in landscape planning. Climatic maps are able to serve as visualization aids from micro to macro level. Through the Geographical Information Systems (GIS) platform, multiple layers of spatial information can be analyzed simultaneously. The use of climatic mapping has become a prominent feature in studies of the outdoor climate (Katzschner et al. 2004; Katzschner and Mülder 2008; Koster 1998; Stocks and Wise 2000). In particular, there has been extensive usage of GIS for the mapping of green spaces (Kamishima et al. 2002; Laing et al. 2006), providing opportunities to propose landscape solutions via GIS mapping techniques. The following section discusses the potential of this methodology through a hypothetical landscape design exercise.

3.2 Plant Selection via Functional Traits

Results from the initial study showed that rooftop greenery can significantly reduce T_{mrt} above the plant canopy. Consequently, a second set of measurements was conducted, this time with the intention of quantifying specific plant traits and their impact on T_{mrt}.

Measurements of T_{mrt} above three rooftop greenery plots were conducted at the same location (Rooftop of the National University of Singapore SDE1). The plants used were Phyllanthus cochinchinensis, Heliconia 'American Dwarf' and Sphagneticola trilobata. Concurrent measurements of plant evapotranspiration rate and shrub albedo were made for a period of 5 months. A detailed description of the study is outlined in Tan et al. (2015). The empirically derived prediction model based on relevant measurement data is as follows:

$$Tmrt_{Plant} = 0.782 Tmrt_{ref} - 200.111 ET - 61.011 SA + 26.937 \tag{17.2}$$

Where,

$Tmrt_{Plant}$ = Mean radiant temperature above rooftop greenery (°C)
$Tmrt_{Ref}$ = Mean radiant temperature above concrete (°C)
ET = Plant evapotranspiration rate ($mm{\cdot}min^{-1}$)
SA = Shrub Albedo

The model can be used to determine the cooling potential of shrubs by assessing their respective evapotranspiration and albedo traits. This enables the landscape designer to select plants based on their ability to reduce temperature. Usage of plants with higher evapotranspiration rates or albedo may be favored in light of their corresponding cooling potential. Since there is lesser chance of reducing T_{mrt} through shade provision by large canopy trees on roof gardens and sky terraces due to structural loading issues, it is important that the shrubs used can help reduce T_{mrt} as much as possible.

4 Landscape Planning Framework Based on Plant Cooling Potential

The proposed landscape design framework is intended to enable landscape designers to evaluate the thermal impact of their design proposals. The workflow is outlined in Fig. 17.6. Digital Elevation Models (DEMs) are the basic spatial layers used for thermal simulation. Raster arithmetic is subsequently employed to deduce the cooling impact of shrubbery. Mean radiant temperature (T_{mrt}) is used to measure plant cooling potential, as this quantity plays a crucial role not only in indoor situations but also outdoors as indicated in several studies which have stressed that outdoor thermal comfort is highly dependent on the short wave and long wave radiation fluxes from the surroundings (Mayer 1993; Mayer and Höppe 1987).

In the hypothetical urban model, four design iterations have been conducted for an area slated to be park space. Simulation is conducted with SOLWEIG (Lindberg and Grimmond 2011) and ARCGis software. In Iteration 1, trees with small canopies (5 m diameter) are placed at locations designated by the landscape planner. In Iteration 2, trees with larger canopies are assumed (15 m diameter) at the same spots. The T_{mrt} reduces drastically near the trees. In Iteration 3, more trees (20 m diameter canopy) are added to areas that are anticipating larger pedestrian flow. As a result, thermal conditions of these areas are shown to have improved significantly. In Iteration 4, thermal effects of shrubbery are factored into the T_{mrt} map via an empirical model based on the cooling effects of plants due to their evapotranspiration rates and albedo values (Eq. 17.2).

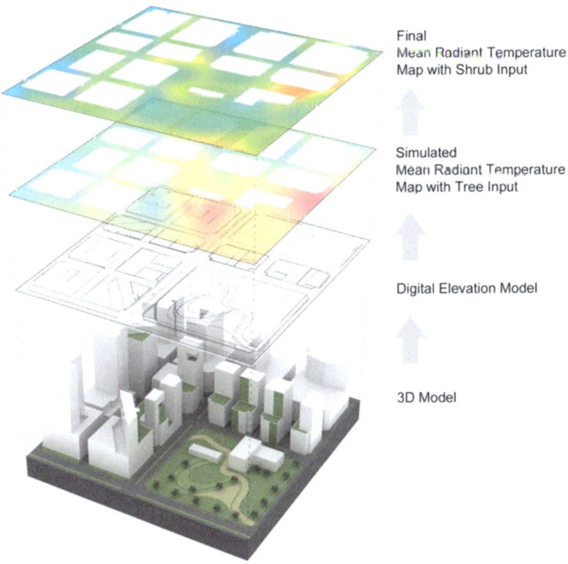

Fig. 17.6 Landscape planning T_{mrt} modelling hierarchy.

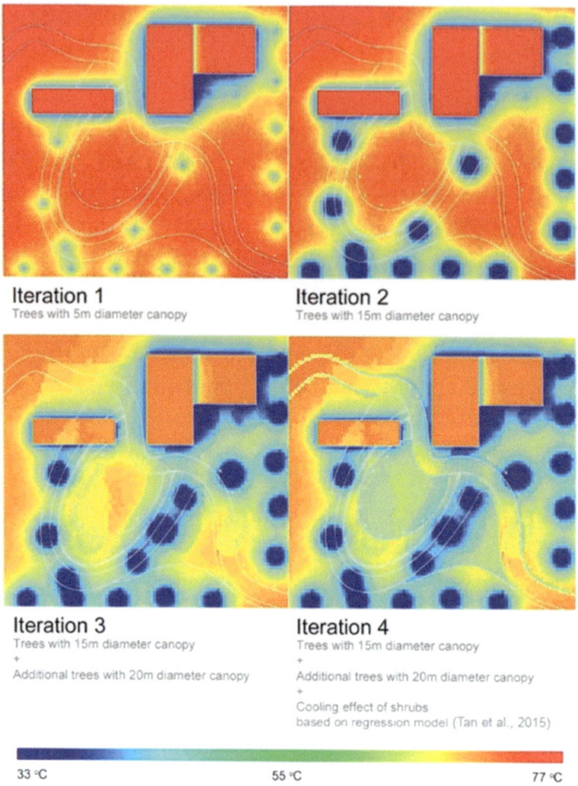

Fig. 17.7 Comparison of T_{mrt} simulation results

Comparison of all four iterations reveals the immense positive impact of tree and shrub allocation using the proposed landscape planning framework (Fig. 17.7).

The proposed landscape design framework allows designers to understand the impact of their choice of plant selection and allocation before eventually committing to a final decision. This can help to minimize undesirable outcomes such as lack of shading provision at prominent locations as well as inadequate light provision for plants due to excessive overshadowing from adjacent buildings.

5 Conclusion

Objective plant selection and placement are important factors in landscape planning. In the above example, scientific objectives are proposed to lend sophistication to the landscape design process. Introduction of thermal simulation in this study has highlighted the importance of context and locality. Adjacent buildings can affect solar exposure significantly, thereby influencing the plant placement process, dispelling the common myth that plants can improve the environment by cooling temperature indiscriminately.

The proposed framework for landscape planning seeks to more effectively realize the cooling effects of greenery as an urban heat mitigation technique and to optimize urban greenery as an ecosystem resource.

References

Akbari H (2002) Shade trees reduce building energy use and CO_2 emissions from power plants. Environ Pollut 116:119–126

Baik JJ, Kwak KH, Park SB, Ryu YH (2012) Effects of building roof greening on air quality in street canyons. Atmos Environ 61:48–55

Beatley T (2000) Green urbanism: learning from European cities. Island Press, Washington, DC

Ca VT, Asaeda T, Abu EM (1998) Reductions in air conditioning energy caused by a nearby park. Energ Buildings 29:83–92

Cameron RWF, Taylor JE, Emmett MR (2014) What's 'cool' in the world of green façades? How plant choice influences the cooling properties of green walls. Build Environ 73(0):198–207

Castleton HF, Stovin V, Beck SBM, Davison JB (2010) Green roofs; building energy savings and the potential for retrofit. Energ Buildings 42:1582–1591

Cheng CY, Cheung KKS, Chu LM (2010) Thermal performance of a vegetated cladding system on facade walls. Build Environ 45:1779–1787

Council of Tree and Landscape Appraisers (1992) Guide for plant appraisal. International Society of Arboriculture, Savoy, III

Gao W (1993) Thermal effects of open space with a green area on urban environment, Part I: a theoritical analysis and its application. J Archit Plann Environ Eng, AIJ, No. 488

Gómez-Muñoz VM, Porta-Gándara M, Fernández J (2010) Effect of tree shades in urban planning in hot-arid climatic regions. Landsc Urban Plan 94(3):149–157

Hough M (1984) City form and natural processes. Croom Helm, London

Humphreys MA (1977) The optimum diameter for a globe thermometer for use indoors. Ann Occup Hyg 20(2):135–140

Jones HG (1992) Plants and microclimate, 2nd edn. Cambridge University Press, Cambridge

Jonsson P (2004) Vegetation as an urban climate control in the subtropical city of Gaborone, Botswana. Int J Climatol 24:1307–1322

Kamishima K, Kohmura K, Mochizuki K (2002), The analysis of greening effects on urban environment using GIS. In: Esri International User Conference. ESRI, San Diego

Katzschner L, Mülder J (2008) Regional climatic mapping as a tool for sustainable development. J Environ Manag 87(2):262–267

Katzschner L, Bosch U, Röttgen M (2004) A methodology for bioclimatic microscale mapping of urban spaces. University of Kassel, Kassel

Koster E (1998) Urban morphology and computers. Urban Morphol 2(1):3–7

Laing RA, Miller D, Davies AM, Scott S (2006) Urban greenspace: the incorporation of environmental values in a decision support system. J Inf Technol Constr 11:177–196

Lindberg F, Grimmond C (2011) The influence of vegetation and building morphology on shadow patterns and mean radiant temperatures in urban areas: model development and evaluation. Theor Appl Climatol 105(3–4):311–323

Mayer H (1993) Urban bioclimatology. Experientia 49(11):957–963

Mayer H, Höppe P (1987) Thermal comfort of man in different urban environments. Theor Appl Climatol 38(1):43–49

McPherson EG (1994) Preserving and restoring urban biodiversity: cooling urban heat islands with sustainable landscapes. In: The ecological city. University of Massachusetts Press, Amherst, pp 151–172
Nikolopoulou M, Lykoudis S (2006) Thermal comfort in outdoor urban spaces: analysis across different European countries. Build Environ 41:1455–1470
Parizotto S, Lamberts R (2011) Investigation of green roof thermal performance in temperate climate: a case study of an experimental building in Florianópolis city, Southern Brazil. Energ Buildings 43:1712–1722
Pérez G, Rincón L, Vila A, González JM, Luisa LF (2011) Green vertical systems for buildings as passive systems for energy savings. Appl Energy 88:4854–4859
Perini K, Ottelé M, Fraaij ALA, Haas EM, Raiteri R (2011) Vertical greening systems and the effect on air flow and temperature on the building envelope. Build Environ 46:2287–2294
Ra JG (2006) Four-year plan of the fourth elected city administration (2006–2010), Seoul, a clean and attractive global city. Seoul Metropolitan Government, Seoul
Saadatian O, Sopian K, Salleh E, Lim CH, Riffat S, Saadatian E, Toudeshki A, Sulaiman MY (2013) A review of energy aspects of green roofs. Renew Sust Energ Rev 23:155–168
Shashua-Bar L, Hoffman ME (2000) Vegetation as a climatic component in the design of an urban street: an empirical model for predicting the cooling effect of urban green areas with trees. Energ Buildings 31:221–235
Simpson JR, McPherson EG (1996) Potential of tree shade for reducing residential energy use in California. J Arboric 22:10–18
Stocks CE, Wise S (2000) The role of GIS in environmental modelling. Geogr Environ Model 4(2):219–235
Tan CL, Wong NH, Jusuf SK (2013) Outdoor mean radiant temperature estimation in the tropical urban environment. Build Environ 64:118–129
Tan CL, Wong NH, Tan PY, Jusuf SK, Chiam ZQ (2015) Impact of plant evapotranspiration rate and shrub albedo on temperature reduction in the tropical outdoor environment. Build Environ 94:206–217
Thorsson S, Lindberg F, Eliasson I, Holmer B (2007) Different methods for estimating the mean radiant temperature in an outdoor urban setting. Int J Climatol 27:1983–1993
Vernon HM (1932) The measurement of radiant temperature in relation to human comfort. J Ind Hyg 14:95–111
Wong NH, Chen Y (2005) Study of green areas and urban heat island in a tropical city. Habitat Int 29:547–558
Wong NH, Chen Y, Ong CL, Sia A (2003) Investigation of thermal benefits of rooftop garden in the tropical environment. Build Environ 38:261–270
Wong NH, Tan AYK, Tan PY, Wong NC (2009) Energy simulation of vertical greenery systems. Energ Buildings 41:1401–1408
Zhao J (2011) Towards sustainable cities in China. In: Analysis and assessment of some Chinese cities in 2008. Springer, New York

Chapter 18
Low Energy Material, High Community Engagement: Ecology of Banana Tree Fibres in Disaster Relief Projects

Mirian Sayuri Vaccari and Lara Leite Barbosa de Senne

Abstract This article describes a research project carried out as part of a larger piece of research into design for emergency relief following flood disasters with a focus on the area of the Ribeira Valley in Brazil. The aim of this paper is to present the results and discuss the ecology of building materials that use banana tree fibre. The article describes the community for whom the work was carried out and gives a detailed overview of the manufacturing process. The research involved an investigation into the production of fibreboard using waste fibre materials from banana cultivation combined with waste paper and a resin. There is also a discussion of the impacts resulting from the manufacture of this material. Based on the comprehension of the issues discussed by collaborative groups of researchers, staff of the local city council, victims of previous floods, artisans and Quilombolas (Brazilian Maroon Communities) and, using paper made of banana tree fibres, a panel wall system is being developed to be used in temporary shelters and toilet cubicles. In order to achieve this objective, a literature review on ecology of building materials is included, in addition to the description of the banana tree fibreboard manufacturing. The outcomes of this article provide information for further design focusing on the development of new components using local materials or vegetable fibres.

Keywords Banana tree fibre • Ecology of building materials • Disaster relief • Quilombola communities

M.S. Vaccari (✉) • L.L.B. de Senne
FAUUSP, Faculty of Architecture and Urbanism, University of São Paulo, São Paulo, Brazil
e-mail: arquitetamirianvaccari@gmail.com

T.H. Karyono et al. (eds.), *Sustainable Building and Built Environments to Mitigate Climate Change in the Tropics*, DOI 10.1007/978-3-319-49601-6_18

1 Introduction

The content of this article is part of a broader research project named "Design for emergency relief: Furniture and Equipment Design for Temporary Shelters for Groups Affected by Disasters related to Rainfall and Flood". This research discusses emergency relief collaborative design and it is developed by the research group NOAH – Núcleo Habitat Sem Fronteiras (Habitat Nucleus without Borders), created and coordinated by Professor Lara Leite Barbosa de Senne, from the Faculty of Architecture and Urbanism of the University of São Paulo, Brazil (FAUUSP). The aim is to design and create solutions and build prototypes that reflect the true demands and aspirations of the people involved in disasters. As part of the collaborative research, studies have been carried out in order to produce building components made from banana tree fibres for capacity building to empower flood victims (Barbosa 2014).

The project started in 2010, and during these 5 years, research, surveys and workshops were implemented to base the objective of the project through participatory design. According to Barbosa (2014), the research involves different educational and research institutions, government officials and local authorities, non-governmental organisations, partner companies and individuals affected by floods in the municipality of Eldorado, located in southern SãoPaulo, in the Ribeira Valley in Brazil. The intervention actions aim to provide resilience for the population, strengthening local development by combining low energy materials with innovative technologies.

In her article entitled "The integration of Traditions, Technologies, Translations and Transformations in Banana Tree Fibre Designs", Barbosa (2014) suggests for the project the promotion and integration of three actions based on the precepts of eco-development stated by Ignacy Sachs (Sachs 1993) in order to adapt and mitigate built environments affected by natural disasters because of climate change.

These three actions comprise designing networks with equality, doing more with less and integrating the community during, not after, the project. The first action relates to the relationships between researchers, consultants, the population and leaders who should live together in the context and be part of the project. The third guarantees access to everyone in the decision-making process, preserving diversity and cultural identity.

The second action, Doing more with less, is the object of study of this article. It relates to the ecological dimension of the project, which should be integrated and coordinated with the project's human dimensions, doing so in an interdisciplinary manner. Thus, it must develop appropriate technologies that allow for the absorption of the best of each ecosystem's human and natural variables. Berge (2000), while defining the ecology of building materials, questions if it is realistic to imagine a technology that functions in line with holistic thoughts while also providing humanity with an acceptable material standard of living. This reality is the core of this paper in the context of building materials for disaster relief: what is the possible role and potential of the use of banana tree fibres in the perspective of the ecology of building materials?

2 The Natural and Human Variables

2.1 The Ribeira Valley

The research on design for disaster relief is focused in the municipality of Eldorado, located in the Ribeira Valley, southeastern Brazil. The Ribeira Valley represents the municipalities with the lowest Human Development Index (HDI) in the states of São Paulo and Paraná, with values lower than these states' average (São Paulo – 0.82 and Paraná – 0.787). The HDI of Eldorado is 0.733 [11] (Serviço Geológico do Brasil 2014).

The climate is humid subtropical, the annual average temperature is 18 °C and annual precipitation can reach up to 4000 millimetres. Altitude varies from 100 to 1000 m. Figure 18.1 shows the location of the city of Eldorado and the Ribeira Valley.

The Ribeira Valley concentrates the biggest continuous area of Atlantic forest in Brazil. According to Fundação SOS Mata Atlântica (1998) cited in Santos (2005), the area represents 2.1 million hectares, equivalent to 21 % of the remaining Atlantic forest in the country, 150,000 ha of restinga (tropical and subtropical moist broadleaf forest) and 17,000 ha of conserved mangrove. It also presents the most important speleological Brazilian heritage being declared Natural Heritage by UNESCO (Santos 2005).

Eldorado is located by the banks of the Iguape River, thus it is often affected by floods when the river can reach levels up to ten meters above average. During the floods, the population moves to temporary shelters, public buildings such as churches, schools and community centres adapted to be used as shelters during the emergency period (Barbosa 2014).

Fig. 18.1 Location of the city of Eldorado and the Ribeira Valley (Source: NOAH)

2.2 *Quilombola Communities*

The Ribeira Valley concentrates most of the Quilombolas communities, living communities of Afro-Brazilians, descendants of enslaved Africans, who organized themselves under local leadership after the end of slavery in 1888. They had their ownership of land recognized only a hundred years after the end of slavery, in the 1988 Brazilian Constitution (Barbosa 2014).

2.3 *Banana Culture and Craftwork*

Agriculture is still the main economic activity in the Ribeira Valley, both 'patronal' agriculture, a Portuguese term, meaning where someone owns the land and other people farm it, and 'familiar' agriculture, a Portuguese term meaning where a family owns and works the land (Resende 2002 cited in Santos 2005, p. 6). Banana is the main produce of the regional agricultural economy followed by beans, rice and corn (Instituto de Terras do Estado de São Paulo 2000, cited in Santos 2005, p.6).

Nowadays, the Quilombolas live from small family subsistence farms where the banana is the main produce that is sold *in natura* or in sweets and is also part of the Quilombolas' daily meals (Santos 2005). They live in Conservation Units (UC) which are land or sea spaces relevant to the maintenance of the ecological equilibrium. They are protected by the government because of their fundamental role in the protection and preservation of the natural environment (Sistema Ambiental Paulista 2014). Before the creation of the Conservation Units (UC) in the Ribeira Valley, the Quilombolas communities made their living, basically, from extractive agriculture or 'coivara' agriculture (in which the forest is destroyed and burnt and the area is cultivated for 3 years). Figure 18.2 shows results of the traditional craftwork techniques to produce building components.

The Quilombolas, traditionally, produced handicraft with raw materials extracted from the forest; however, they did not attribute the word "craftwork" to the objects they produced. They did not produce decorative objects, but everyday utensils such as tipiti (a type of cassava masher), sieves and wooden shovels. The use of the term "craftwork" was adopted when the School of Agriculture from the University of São Paulo (ESALQ) started banana tree fibres handicraft capacity-building courses and the community recognized their craftwork as having commercial appeal. With the work of the University of São Paulo, the use of banana culture residues was reorganized when the residue was not considered waste but a potential resource to be explored by the communities (Santos 2005). In 2014, the research group NOAH organised a workshop for experimentation with banana tree fibres. The aim was essentially to collect various information with the social groups, which are vulnerable to the impacts of events related to rain and flooding (Barbosa et al. 2014a).

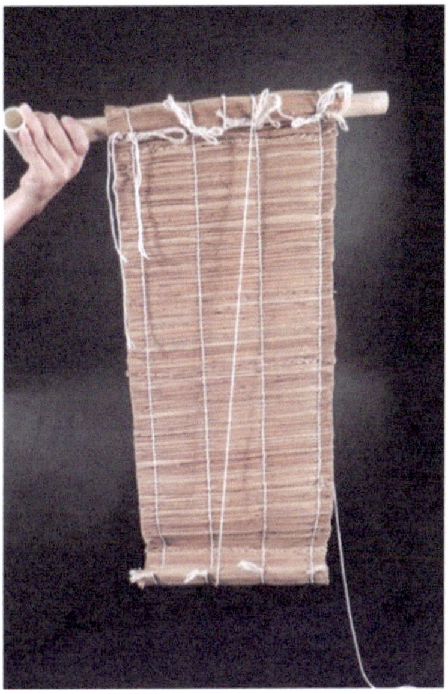

Fig. 18.2 Results of the III Workshop NOAH, when the group and the Quilombolas used traditional craftwork techniques to produce building components (Source: NOAH)

3 The Ecology of Building Materials

Through the natural and human variables explained above and the aspects of the ecology of building materials, the goal of this paper is to analyse each feature of the use of banana tree fibres in the context of the production of building components by the victims of flooding. The attributes to be studied in the ecology of building materials, according to Berge (2000) are detailed as follows:

- *Work*: The methods used to produce the building component. How production takes place and can take place.
- *Raw materials*: Occurrence of material resources, their nature, distribution and potential for re-use.
- *Energy:* The energy consumed when producing and transporting the materials, and their durability.
- *Pollution:* Pollution during production, use and demolition, the chemical fingerprint of each different material.

In order to assess the ecology of the banana tree fibreboard which is the subject of this article, the manufacturing of the boards is explained below. The description of the process is based on project reports of the group NOAH,of which Professor Lara Leite Barbosa is the coordinator and Mirian Sayuri Vaccari was the research monitor in 2013/2014. The reports, Barbosa and Vaccari 2014, Barbosa et al. 2014a, b, c and Barbosa et al. 2013, are results of two subprojects of the group:

- Elaboration of a manual for the production of building components using vegetable fibres: This project's aim is capacitation building by producing a manual

for victims of flooding in the Ribeira Valley to enable the community to produce building components with vegetable fibres
- The APIS Project: Emergency bathrooms for use following rain-related disasters: a partnership of Architecture for Humanity and Alcoa Foundation. For this project, the group will design and build temporary and portable ablution blocks for public use in post-disaster situations.

Both projects are a result of research, field trips and workshops organized by the research group NOAH. The aim was to collect various information with the social groups vulnerable to the impacts of flooding and from Eldorado's city council employees, especially from the Agriculture Department and public employees or NGOs that act in the studied location.

3.1 Manufacturing the Boards

The boards are manufactured in three phases described below: extraction of the raw material (banana tree fibre) from the forest, manufacturing of the banana tree fibre paper and manufacturing the boards.

3.1.1 Extraction and Drying of Banana Tree Fibre

For the extraction of fibres from the banana trees, the traditional process already used by the Quilombolas is maintained. Firstly, the tree is cut, in the day of waning moon, as it is believed by the Quilombolas that the contamination by termites is thereby avoided. According to the Quilombolas, the banana tree must be at least 6 months old so that the false stem is sufficiently grown for the cut and for its fibre use. When the aim is also to produce bananas, one should wait around 9 months for the banana to grow and another 6 months more for the maturation, totalling almost a year and a half (Figs. 18.3 and 18.4).

3.1.2 Banana Tree Fibre Papermaking

Paper made of banana tree fibres and recycled common paper follows basically the same steps of paper recycling - trituration, cooking, mixing, adding water to the mixture, filtering and drying. The paper is made by Eldorado residents in the municipality's factory.

Firstly, the material extracted from the banana tree is triturated. The first raw material used for papermaking was the false stem of the banana tree. However, being a waste material from banana culture and handicraft, the peduncle (the part which connects the bananas to the false stem), was substituted for the false stem as it is generally acquired free from the producers.

Fig. 18.3 Cutting the banana tree from the forest and peeling the false stem (Source: NOAH)

Fig. 18.4 Separating the false stem layers for craftwork and hanging the layers to dry (Source: NOAH)

In the current process, approximately 40 kg of peduncle is used every time when the process occurs. The triturated material is cooked in a cauldron where it is boiled with a small amount of ashes, around 2 kg for the 40 kg of peduncle usually used. After cooking, the mixture is placed in a papermaking mixer. In this phase of the process, an amount of recovered paper is added in a proportion of 5 kg for the 40 kg of peduncle used. Water is added and the mixture then has a proportion of 95 % of water and 5 % of vegetable fibres.

This proportion can be changed and this alteration affects directly the quality of the final product. Mixtures with higher fibre concentration generate thicker and more fibrous papers/boards. This mixture is placed in tanks where it will "rest" for some period. Finally, the mixture is deposited in the mould and deckle for the final product. The measurements of the mould are around 1 m × 0.80 m, varying accordingly with the different sizes needed. When is it deposited, the material dries through the filtering process (without a press), resulting in a sheet of banana tree fibre paper.

3.1.3 Manufacturing the Boards

For the research, empirical tests were made in the workshop in the city of Eldorado and LAME, the laboratory of models and testing at the University of São Paulo. The steps in manufacturing the banana fibre boards were cutting the paper, gluing the plies and waterproofing. Searching for materials with lower environmental impact for gluing the plies, tests were made with cassava glue, wheat glue, CPA castor oil resin, eco resin Metalatex and beeswax. The cassava and wheat glues were made in the laboratory and the castor oil resin and the eco resin were found in the market (Fig. 18.5).

After finding that problems with mould resulted from the use of glues based on natural materials in places with high humidity, we started using the Imperveg glue. According to the manufacturer, the IMPERVEG UG 132 A is a waterproofing resin based on vegetable polyurethane (originated from castor oil), bi-component, 100 % solid (solvent free), which makes a monolithic membrane on the applied surface (Brazil. Imperveg Polímeros Indústria e Comércio Ltda. 2014). The membrane presents physical-chemical stability, elasticity, waterproofing and adherence to porous materials such as concrete, mortar and wood.

Still in the experimentation phase, the Imperveg is being used. The boards are currently being made of five paper layers, and each layer is composed of a total of three sheets together. The size of each banana fibre paper sheet is approximately 70 cm × 96 cm, so that the board final size after trimming is 90 cm × 200 cm and 3 mm thick.

Fig. 18.5 Board layering, board before trimming and waterproofing test (Source: NOAH)

3.2 Work: Extracting Banana Tree Fibres and Producing the Panels Locally

According to Berge (2000), there are basically three ways of manufacturing a product:

- It can be manufactured by the user, based on personal needs or on the local cultural heritage;
- It can be manufactured by a craftsman who has developed a method of manufacture through experience;
- It can be manufactured by an engineer who directly or indirectly, through electronics, tells the worker which steps to take.

Berge (2000) states that the first two methods share a common factor – the spirit of the product and the hand that produces it belong to the same person. The primary relationship (relationship between producer and consumer in worker-controlled production) has positive effects for the consumer, the manufacturer and the worker. In this paper case study, when the Quilombolas produce the banana tree fibre boards they can concentrate either 3 or 2 functions in the chain -consumer, manufacture and worker – depending if they are producing for their own use in disaster situations or if they are producing the boards for commerce. The primary relationship can lead to a better product, the guarantee that the resources will be used responsibly, producing less pollution, less bureaucracy (as in most cases there is a feeling of solidarity in the primary relationship) and flexibility (spontaneity in the production process), safer places to work and meaningful work.

3.3 Raw Materials

The basic materials of the banana tree fibreboards are banana tree peduncles, recycled paper and resin. The banana tree peduncles as well as the recycled paper are unused resources and waste products and renewable resources, mentioned by Berge (2000) as two of the actions for reducing the use of raw materials in the production process. The other actions are exploitation of smaller sources of raw material and increased recycling of waste products during production. Figure 18.6 shows the banana tree.

For the exploitation of smaller sources, the banana tree fibre extraction is based on the Quilombolas´ family farms and waste products from banana culture. Hiroce (cited in Santos 2005) estimates that the false stem (from where the banana tree fibre can be extracted) is constituted of approximately 92 % water and 3 % fibre, meaning that a 40-kilogram stem would provide 1.2 kilograms of fibre. Thus, after the fruit is collected, from one hectare of banana trees, with 2000 banana trees (spaced at 2 × 2 m), around 2.4 tonnes of fibres are produced.

The disadvantages of the banana tree fibreboard in terms of reducing the use of raw materials are the use of resin (which will be mentioned in Sect. 3.4 below) and

Fig. 18.6 Banana tree and the Ribeira River (Source: NOAH)

recycling of waste products during production as well as recyclability or reuse of the boards.

During the production of the boards, the process that produces the most waste products is papermaking, which produces a lot of wastewater. In further development stages of the project, wastewater should be a concern in terms of reducing the use of raw materials.

When addressing the recyclability of paper, Schönwalder (cited by Vaccari 2008), mentions that the paper itself is not the most important issue. The fillers (banana tree fibre) and adhesives (Imperveg resin) which are necessary for the strength and impregnation of the construction element, bring the subsequent recycling into question. Finding the most ecological solution of this issue is essential for the use of banana tree fibre paper in architecture.

3.4 Energy

In this item, only the primary energy consumption of banana tree fibreboard manufacturing will be analysed. The energy consumption during building, use and demolition will not be taken into consideration for this analysis of the banana tree fibreboards.

According to Berge (2000), the primary energy consumption is usually about 80 % of the total energy input in a material and is divided into the direct energy consumption in extraction of raw materials and the production processes; secondary consumption in the manufacturing process and energy in transport of the necessary raw and processed materials.

The energy used in the extraction of raw materials, both the banana peduncles and the recovered paper, and the energy in transport are really small in the process. The Quilombolas go to the forest by foot and carry the false stem and peduncles on their back. As for the recovered paper, the Quilombolas use recovered paper from the Eldorado city council.

The problem in the use of energy is the energy used to transport the resin and secondary consumption during papermaking that can be an energy intensive process. Still, Mieli (cited by Vaccari 2008) states that the use of recovered fibres to produce paper can use 23–74 % less energy, reduce air pollution by 25 %, reduce water pollution by 65 % and use 58 % less water if compared to the use of virgin fibres for paper production (for instance, if we used banana tree cellulose to make the paper for the boards). Thus, adding recovered fibres to the banana tree papermaking process may reduce the use of energy and pollution.

3.5 Pollution

When it comes to pollution, the problems can be referred to in terms of energy pollution and material pollution (Berge 2000). As the energy pollution relates strongly to the primary energy consumption and the source of energy (and in Brazil it is mostly hydroelectricity), pollution in banana tree fibre board manufacturing, is focused on papermaking which still uses less energy because of the use of recovered fibres.

Material pollution refers mainly to pollutants in air, earth and water from the material itself and the constituents of the material when being worked, in use and during decay. The resin Imperveg used to glue the banana tree fibre paper plies is the material that can cause the biggest environmental impact. The manufacturer (Brazil. Imperveg Polímeros Indústria e Comércio Ltda. 2014) claims that it is a solid material (solvent free), and it can be applied in confined spaces, because it does not release toxic vapours. It meets the ordinance that defines patterns of water potability for human consumption, and it can be used without restrictions as waterproofing for potable water tanks. Even if the manufacturer claims that the product is non-toxic, studies together with chemistry professionals should be made so as to diminish still further the environmental impact of the banana tree fibreboard.

4 Final Considerations

Principles for an ecological building industry include the following (Berge 2000):

- The technological realm is moved closer to the worker and user, and manufacturing takes place in smaller units near to the area where the products will be used;

- The use of raw materials based on renewable resources or rich reserves, products are easily recycled and are economic in terms of materials during construction;
- Priority is given to production methods that use less energy and more sustainable materials, and transport distances are reduced to a minimum.
- Polluting industrial processes and materials are avoided, and energy based on fossil fuels reduced to a minimum.

From the assessment in this paper, it can be concluded that the banana tree fibreboard follows the basic principles of the ecological building industry. The use of banana tree fibre for producing boards can be a technology that functions in line with holistic thoughts while also providing people with an acceptable material standard of living (Berge 2000). However, production of the banana tree fibreboards is still in an experimental phase. Thus, issues regarding the pollution and the reuse or recycling of the banana tree fibreboard after demolition should be analysed. In additionits thermal properties, durability and stability will be studied in partnership with IPT (Institute for Technological Research). As long as the research improves, the Quilombolas may produce an ecological building material from traditional and natural materials, which may provide them greater resilience to disasters in the future.

References

Barbosa LL (2014) The integration of traditions, technologies, translations and transformations in banana tree fibre design. In: International committee on design history and design studies – ICDHS, 9th conference of the ICDHS International Committee for Design History and Design Studies. Aveiro, Portugal, 09–11 July 2014, Aveiro: UA Editora

Barbosa LL, Vaccari MS (2014) Building elements development and preparation of the tests of the fibres components for the APIS Project- Emergency bathrooms for use following rain-related disasters. Report. NOAH/FAU-USP, São Paulo

Barbosa LL, Cabral GHM, Delanez LG, Vaccari MS (2013) Field trip report. Report. NOAH/FAU-USP, São Paulo

Barbosa LL, Cabral GHM, Delanez LG, Vaccari MS (2014a) Third workshop NOAH – Participatory activities for the experimentation with vegetable fibre. Report. NOAH/ FAU-USP, São Paulo

Barbosa LL, Cordeiro AV, Leone TG (2014b) Report of activities of October. Report. NOAH/FAU-USP, São Paulo

Barbosa LL, Delanez LG, Vaccari MS (2014c) Banana tree fibreboards: one month after the workshop. Report. NOAH/FAU-USP, São Paulo

Berge B (2000) The ecology of building materials. Architectural Press, Oxford

Brazil. Imperveg Polímeros Indústria e Comércio Ltda (2014) Imperveg UG 132A. Imperveg Polímeros Indústria e Comércio Ltda, Aguaí

Sachs I (1993) Estratégias de transição para o século XXI. In: Bursztyn M (ed) Para Pensar o Desenvolvimento Sustentável. Brasiliense, São Paulo

Santos KMP (2005) A atividade artesanal com fibra de bananeira em comunidades quilombolas do Vale do Ribeira (SP). Master in ecology and agrosystems. University of São Paulo

Serviço Geológico do Brasil (2014) Projeto Geoparque Alto Vale do Ribeira. Justificativas Socioeconômicas. [Online]. CPRM, Brasília. Available at: http://www.cprm.gov.br/Geo_Site/socioeconomico.htm. Accessed 5 Dec 2014

Sistema Ambiental Paulista (2014) Parques e Unidades de Conservação [Online]. Governo do Estado de São Paulo, São Paulo. Available at: http://www.ambiente.sp.gov.br/ambiente/parques-e-unidades-de-conservacao/. Accessed 5 Dec 2014

Vaccari MS (2008) Environmental assessment of cardboard as a building material. MSc in Energy Efficient and Sustainable Building. Oxford Brookes University

MIX
Papier aus verantwortungsvollen Quellen
Paper from responsible sources
FSC® C105338

If you have any concerns about our products,
you can contact us on
ProductSafety@springernature.com

In case Publisher is established outside the EU,
the EU authorized representative is:
Springer Nature Customer Service Center GmbH
Europaplatz 3, 69115 Heidelberg, Germany

Printed by Libri Plureos GmbH
in Hamburg, Germany